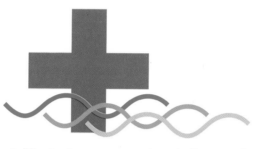

All Stings Considered

All Stings Considered

First Aid and Medical Treatment of Hawai'i's Marine Injuries

Craig Thomas, M.D.
and
Susan Scott

A Latitude 20 Book
University of Hawai'i Press
Honolulu

Library of Congress Cataloging-in-Publication Data
Thomas, Craig, 1952–
    All stings considered: first aid and medical treatment of Hawai'i's marine
injuries / Craig Thomas and Susan Scott.
        p.   cm.
    Includes bibliographical references and index.
    ISBN 0–8248–1900–4 (pbk.: alk. paper)
    1. Aquatic sports injuries--Hawaii. 2. Medicine, Naval--Hawaii.
3. Poisonous fishes--Hawaii. 4. Marine invertebrates--Hawaii.
I. Scott, Susan, 1948– . II. Title.
RC1220.A65T48   1997
616.9'8022'09969--dc20   96–44254
                                                          CIP

Designed by Custom Editorial Productions, Inc.

To editor, surfer, and friend Greg Ambrose,
whose vision inspired this book

CONTENTS

FOREWORD

Craig Thomas, M.D., an emergency room physician, and Susan Scott, a registered nurse and marine science writer, have created a superb book in *All Stings Considered*. Their years of clinical experience and their participation in activities on, under, and around Hawai'i's waters make them *the* authorities on marine injuries in Hawai'i.

This compendium, well researched and well written, is easy for the general reader to understand and will be a handy reference for all water enthusiasts. It is also an excellent medical text for the health professional. No other book specifically addresses the Hawai'i aquatic environment. The many articles on specific marine injuries that exist in the medical literature are difficult for the lay reader to obtain and understand.

Every emergency room and poison control center should have a copy of this book at hand. Paddling, sailing, and swimming clubs, along with high school and college athletic coaches, will use it for accident prevention and quick treatment. First responders, emergency physicians and staff, pediatricians, family practitioners, internists, and dermatologists will find it valuable as a speedy reference.

In their first joint publication effort, the authors have compiled a wealth of marine information. This book will be a *must read* for the thousands of people who enjoy the ocean in Hawai'i. It is a winner.

Norman Goldstein, M.D.

Clinical Professor, Medicine (Dermatology), John A. Burns School of Medicine, University of Hawai'i

Editor, *Hawai'i Medical Journal*

PREFACE

Our intention in this book is to allay fears, not generate them. Knowing how to avoid marine injuries, and how to treat them when they occur, decreases anxiety and allows more people to enjoy our magnificent ocean playground safely.

Our goal is to provide clear, useful information for a wide range of people. Most of the book is in everyday language. We chose to use familiar terms like "emergency room" and ER even though some people in the medical community are now using "emergency department" and ED. We use medical terms in describing Advanced Medical Treatment techniques because medical workers use unique terminology administering these treatments.

All Stings Considered discusses injuries from marine plants and animals in Hawai'i. Gathering information for this limited geographical area presented a problem; we discovered that research on Hawai'i species often was sparse or nonexistent. We occasionally found that similar species had been studied in other parts of the world, but not here. Our approach to this uncertainty has been to use Hawai'i-specific literature whenever possible and to apply data from other areas whenever pertinent. When the literature does not contain any information on a subject, or when it is not applicable to Hawai'i, we say so in the text.

Like all regions, Hawai'i has popular home remedies for marine injuries, few of them scientifically tested. Although some folk cures may indeed help in pain control or healing, others, such as the application of urine or heat to jellyfish stings, may be harmful. Whenever possible, we base our recommended treatments on published, controlled scientific studies. The sciences of medicine and biology, however, are constantly evolving. Researchers publish new data, ships' ballast waters and hulls introduce new species to the waters in Hawai'i, and people have experiences that redefine old ideas and practices. We plan to track these changes and update our book as necessary.

For certain injuries, we ask readers to discard their old beliefs and switch to new treatments. This can be difficult, both for lay people and for professionals. Often, years pass before new information from published studies changes standard medical practice.

One subject we do not discuss is disease transmission through pollution of Hawai'i's ocean waters. This is not an omission. Fifty years of public health monitoring here has revealed no significant illnesses from this source.

We urge interested readers to consult our sources and to contact us if they disagree with any of our conclusions. Readers can notify us about new incidents, new data, or any errors we may have made.

ACCIDENT READINESS

Skills and Equipment for Safe Ocean Outings

Everyone who participates in ocean activities of any kind should know the following basic safety techniques.

- ♦ Know how to get help. Call 911 from shore or from a cellular phone in coastal waters. Contact the Coast Guard by dialing *USCG on a cellular telephone, or by VHF or HF radio (the Coast Guard does not monitor CB radio).
- ♦ Learn to swim.
- ♦ Take a first-aid course.
- ♦ Become certified in cardiopulmonary resuscitation (CPR).
- ♦ All scuba divers should be dive certified.
- ♦ Dive leaders should take the Divers Alert Network's (DAN) oxygen therapy course.
- ♦ Do not drink alcohol during ocean outings. Alcohol impairment greatly increases the risk of marine injuries and deaths.
- ♦ Keep your tetanus (lockjaw) immunization up to date. People already immunized need a booster at least every ten years.
- ♦ Wear a hat, sunglasses, and waterproof sunscreen (SPF 30 or higher) during all ocean outings.

Oceangoers should assemble and carry a first-aid kit that includes the following items.

- ♦ Nonsterile latex gloves (for protection from blood and stinging tentacles).
- ♦ Vinegar (for box jellyfish, fireworm, and sponge stings).
- ♦ Sticky tape (for bandaging and for removing fireworm and sponge bristles).
- ♦ Tweezers (for removing tentacles and sea urchin spines).
- ♦ A mild antiseptic (soap, povidone-iodine, hydrogen peroxide).
- ♦ Antibiotic ointment or cream (for cuts). Avoid neomycin, an antibiotic ointment that frequently irritates skin.
- ♦ Hydrocortisone cream, gel, lotion, or ointment (for rashes).
- ♦ Alcohol or boric acid ear-drying drops (to prevent swimmer's ear).
- ♦ Gauze (for scrubbing or bandaging).

- Elastic bandage (for pressure dressing or splinting).
- Hot packs (for fish stings).
- Cold packs (for tentacle stings).
- Antihistamine tablets (for mild to moderate allergic reactions).
- Automatic epinephrine (adrenaline) injector (for life-threatening allergic reactions–prescription required).
- Oxygen (for divers trained in DAN's oxygen therapy course).

People traveling far offshore and anyone in areas remote from medical help should carry several specific medications discussed in different sections of this book. See your doctor for prescriptions and instructions on the use of these drugs. Prescription drugs are indicated by an asterisk (*).

- Mupirocin* ointment (antibiotic for superficial infections).
- Cephalexin*, 500-milligram tablets (antibiotic for mild to moderate infections).
- Doxycycline*, 100-milligram tablets (antibiotic to add to cephalexin for persistent, worsening, or blistering infections).
- Ciprofloxacin*, 750-milligram tablets (an alternative to doxycycline for severe infections).
- Ceftriaxone*, 2 grams injectable (for life-threatening infections; use with doxycycline or ciprofloxacin).
- Ofloxacin* ophthalmic drops (for swimmer's ear infections).
- Meclizine, 25-milligram tablets (for seasickness).
- Compazine*, 25-milligram suppositories (for severe seasickness).
- Pseudoephedrine (for ear or sinus squeeze and seasickness).

QUICK-LOOK GUIDE TO HAWAI'I'S COMMON MARINE INJURIES

Hazard	Frequency	Dangerous?	First Aid
Anemone stings	Rare	Painful blisters	Rinse with water; use ice for pain relief
Barracuda and eel bites	Occasional	Lacerations often require suturing	Direct pressure for bleeding; scrub
Box jellyfish stings	800 cases/year on leeward side	Painful; no Hawai'i deaths	Vinegar rinse; pick off tentacles; then ice
Bristleworm stings	Occasional	Painful rash	Remove bristles with tape; soak in vinegar
Ciguatera fish poisoning	About 80 Hawai'i cases/year	No Hawai'i deaths	Stop others from eating the fish
Cone snail stings	Rare	Potentially fatal; no Hawai'i deaths	Pressure bandage
Coral cuts	Common	Frequent infections	Scrub well; apply antibiotic ointment
Crown-of-thorns seastar stings	Occasional	Painful punctures	Pull out spines
Drowning	Hawai'i has highest incidence in United States	70 deaths/year	Start CPR; call 911
Hydroid stings	Occasional	Painful rash	Rinse with water; apply ice
Infected cuts	Common	Bloodborne infection is life threatening	Scrub; use topical or oral antibiotics
Portuguese man-of-war-stings	6,500/year on windward side	Painful; no Hawai'i deaths	Pick off stinger; wash with water, then ice
Ray stings	Rare	Painful, slow-healing wounds	Scrub; rinse; remove stinger; apply heat
Scombroid fish poisoning	About 50 Hawai'i cases/year	No Hawai'i deaths	Antihistamines; stop others from eating the fish
Scorpionfish stings	Occasional	Painful; no Hawai'i deaths	Scrub; rinse; apply heat

(Continued on p. 4.)

Hazard	Frequency	Dangerous?	First Aid
Sea urchin *(wana)* punctures	Common	Painful punctures	Pull out large spines; let small spines dissolve
Shark bites	Hawai'i averages 1 attack every 2 years	About ⅓ of shark bites are fatal	Use direct pressure for bleeding; call 911
Spinal injury	129 beachgoers in 1995	Paralysis or death	Immobilize; call 911
Sponge stings	Occasional	Painful rash	Remove bristles with tape; soak in vinegar
Stinging limu	Occasional	Itchy rash	Shower; apply steroid cream
Sunburn	Common	Increases skin cancer risk	Use sunscreen, visor, protective clothing
Swimmer's ear	Common	Causes ear canal damage	Dry ear canals after swimming
Zoanthid *(limu-make-o-Hāna)* poisoning	Rare	Dangerous if eaten or penetrates skin	Keep limu out of cuts or mouth; call 911

Use Vinegar	Apply Ice	Apply Heat
Box jellyfish stings Fireworm stings Sponge stings	Anemone stings Box jellyfish stings Hydroid stings Portuguese man-of-war stings	Leatherback fish stings Ray stings Scorpionfish stings Squirrelfish stings

Part 1

Bites, Cuts, and Stings

In the Red Sea, the 3-inch-tall anemone *Triactis producta* produces a powerful sting. This species grows only to about 1/2 inch tall and 1/2 inch wide in Hawai'i. Kāhala. (Colin J. Lau)

ANEMONE STINGS

Sea anemones, which look like fleshy flowers, are relatives of jellyfish, Portuguese man-of-war, corals, and hydroids. Most animals of this group catch their prey with tentacles, paralyzing it with stinging cells called nematocysts.

Each sea anemone is a solitary animal with a soft, vaselike body. At the top of the base is an oral disk with a slit-shaped mouth in the center. At one or both sides of the mouth are grooves, pumping water in and out of the body. This circulating water is how the animal breathes, and it also acts as a fluid skeleton, supporting the creature.

Anemones are carnivores, feeding on animal plankton, invertebrates, and sometimes even fish. After stinging a prey, the anemone transfers the paralyzed creature to its mouth and swallows it whole.

At least twenty kinds of anemones, ranging from about 1/4 to 8 inches tall and from about 1/8 to 6 inches wide, live in Hawai'i. Their tentacles number from sixteen to about four hundred, depending on the species. Five of Hawai'i's anemones are found only in Hawaiian waters.

Anemones live in sand, under rocks, and in crevices. Some spend their entire lives in one place; others move around. Two types of crabs in Hawai'i carry small anemones in their claws for defense. Other anemones live attached to hermit crab shells.

Some species of fish, shrimp, crabs, and brittle stars (sea stars with flexible arms) live among anemone tentacles without getting stung.

Mechanism of Injury

Anemones have nematocysts on their waving tentacles and sting on contact. The nematocysts discharge toxins, which can impair nerve and heart function and cause tissue and blood breakdown. Eating an anemone can cause severe illness.

Box jellyfish and Portuguese man-of-war are the most common sources of nematocyst stings in Hawai'i. For a detailed discussion of nematocysts, see Box Jellyfish Stings and Portuguese Man-of-War Stings.

Incidence

Hawai'i's anemones are small and scarce compared to some tropical areas, so stings here are not common. Hawai'i hosts one sea anemone, *Triactis producta,* which is known to produce severe stings in the Red Sea, where the animal grows to about 3 inches tall and 1 1/2 inches wide. In Hawai'i, this anemone grows to only about 1/2 inch tall and 1/2 inch wide. Hawai'i's small crab *Lybia edmondsoni* often carries *Triactis producta* in its claws for defense. One local author reports having captured and handled the tiny crabs and their anemones without suffering any ill effects.

In the Adriatic Sea, the sea anemone *Anemonia sulcata* commonly stings swimmers. Hawai'i hosts its close relative *Anemonia mutabilis,* which is about 1 1/2 inches wide and 1 inch tall. It too can cause stings. At least two other Hawai'i anemones have produced significant stings. One was probably *Telmatacis decora,* which stung a child in a Moloka'i tide pool. Another was *Stoichactis* sp., which stung an aquarium worker during tank cleaning.

Prevention

Do not touch (or eat) any sea creature that looks like a flower and has tentacles. Divers should wear wet suits and gloves to prevent accidental encounters. Some sea anemones can sting through Lycra suits.[1]

Signs and Symptoms

People usually compare an anemone sting to a bee sting. Skin reactions vary according to the individual and the species. Most commonly, a pale hive appears, encircled by a red halo. The affected area soon swells and looks bruised. In most cases, symptoms disappear within forty-eight hours.

If the sting is severe, localized bleeding, blisters, skin ulcers, and infection may result. The area can remain sensitive for five weeks, and may be permanently discolored or scarred.

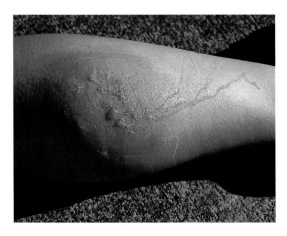

This victim was scuba diving on Oʻahu's Waiʻanae coast when her leg brushed against a rock or coral head. The patient did not see what stung her, but she experienced an immediate burning sensation. Several hours later, the sting, probably from an anemone, looked like this. (Craig Thomas, M.D., and Susan Scott)

Sometimes, lymph nodes in the vicinity of the sting may swell. This is usually a sign of the toxic effect of the venom, or it may be an early sign of infection. Stings to the eyes can cause irritation, pain, swelling, tearing, blurred vision, or light sensitivity.

Overall illness from anemone stings is rare; symptoms include fever, chills, sleepiness, weakness, nausea, vomiting, and dizziness.

 **First Aid**

Pick off any visible tentacles. Rinse the sting thoroughly with salt or fresh water to remove any adhering nematocysts. Apply ice to the skin for pain relief. For persistent itching or skin rash, try 1 percent hydrocortisone ointment four times a day, and one or two 25-milligram diphenhydramine (Benadryl) tablets every 6 hours.[2] These drugs are sold without prescription. Diphenhydramine may cause drowsiness; do not drive, swim, or surf after taking this medication.

Nearly every substance imaginable has been applied to anemone stings throughout the world, including manure, mustard, and figs. No studies indicate that any of these neutralize the venom.[3] Some may be harmful.

Irrigate eye stings with copious amounts of room temperature tap water for at least fifteen minutes. If vision blurs or the eyes continue to tear, hurt, and swell or are light sensitive after irrigating, see a doctor.

If a red streak develops between any swollen lymph nodes and the sting site, or if either area becomes red, warm, and tender, see a doctor immediately. For information about signs of infection, see Part 2, *Staph, Strep,* and General Wound Care.

Rarely, anemone stings cause seizures and death. Eating an anemone can be fatal. If pain persists, the skin rash worsens, allergic

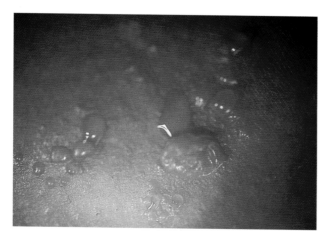

A close-up of the sting on the evening of the incident. A lymph node in the victim's groin area swelled to the size of a grape. (Craig Thomas, M.D., and Susan Scott)

symptoms occur, or a feeling of generalized illness develops, get medical help quickly.

Advanced Medical Treatment

No specific antidote or clinically useful diagnostic tests exist for anemone stings.

Anemone stings may need debriding and antibiotics for secondary infections. (See Part 2, *Staph, Strep,* and General Wound Care.) Healing of anemone stings is slower than Portuguese man-of-war stings.

Some nematocyst stings have recurrent dermatitis with itching but no pain. Therapy is unproven. For persistent rash and itching, try oral steroids for several days.

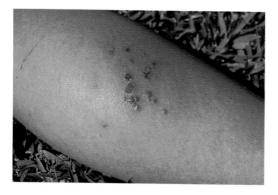

Two days after the sting, the blisters had begun to dry up and the swollen lymph node had disappeared. Two months later, the lesions were healed, but the skin remained discolored. (Craig Thomas, M.D., and Susan Scott)

Peripheral nerve deficit (motor and sensory) after an anemone sting has been reported. Symptoms developed hours after the sting, affected nerves near the sting site, and were not associated with impaired circulation. Motor and sensory symptoms resolved over months without treatment.[4]

No literature exists on anemone stings to the eye. Eye stings from the sea nettle, a jellyfish common in Chesapeake Bay, are usually self-limited, disappearing in twenty-four to forty-eight hours. One report recommends using topical steroids and cycloplegic drops for these stings. Some victims have persistent dilated pupils with blurred vision, particularly for nearby objects. Initially, intraocular pressure is elevated. This pressure remains high in some patients, producing glaucoma in the affected eye.[5]

For anemone ingestion, lavage if the patient arrives within one hour of ingestion. Avoid emetics because of possible seizures. Give charcoal and treat supportively. Complications can include hypotension, seizures, respiratory failure, hepatic failure, and coma. One anemone sting patient in the Caribbean died from hepatic necrosis despite receiving a liver and kidney transplant.[6]

BARRACUDA (KĀKŪ) BITES

Barracuda hunt other fish, striking with great bursts of speed. Unlike some fish predators, barracuda cannot expand their mouths to swallow large fish whole. To eat large prey, barracuda slash them to pieces with remarkably sharp teeth. A large barracuda can cut a mature parrotfish in two with a single bite.

Hawai'i hosts two species of barracuda, the great barracuda, or *kākū (Sphyraena barracuda),* and Heller's barracuda, or *kawele'ā (Sphyraena helleri).* Great barracuda are active during the day. Juveniles live in sheltered inner reefs and harbors. Adults range from murky inner harbors to the open ocean. Individuals can grow to nearly 6 feet long and weigh up to 100 pounds. Heller's barracuda are active at night, forming schools near the reef during the day. This smaller species grows to about 2 feet long.

Mechanism of Injury

Barracuda have two parallel rows of extremely sharp cutting teeth in both the upper and lower jaws. The teeth can inflict deep, slashing cuts,

This 6-foot great barracuda, named Barry by *Atlantis* submarine divers, often hovered near the sub in Kona waters. The barracuda became a pet to the divers, who sometimes fed it bits of fish held at the end of a stick. No accidental bites occurred. (Dean Sensui)

often causing nerve and tendon damage, and sometimes severing large blood vessels. Barracuda teeth can break off inside wounds.

Incidence

Barracuda attacked two Maui fishermen in the 1960s. A 6-foot barracuda slashed the leg of one man who was throw-net fishing. The resulting injury to his left foot and leg required five hours of surgery. The second man needed 255 stitches to repair arm wounds. In the late 1980s, a barracuda attacked a scuba diver at an isolated rock offshore from Kailua, O'ahu. The resulting wound was minor. In Kona, on the island of Hawai'i, doctors treated two women in separate incidents for barracuda bites to the scalp. Both women were wearing shiny barrettes in their hair at the time of the attack. One woman, attacked in the 1990s, required surgery to remove embedded barracuda teeth.

Of the twenty-two barracuda species throughout the world, the great barracuda is the only one known to attack humans. The risk of being bitten by this fish, though, is extremely low. In a study of twenty-nine reported barracuda attacks in the United States between 1873 and 1963 (ninety years), only nineteen were confirmed.

Probably the greatest threat barracuda pose to humans is ciguatera, a poison these fish sometimes carry in their flesh. Ciguatera poisoning comes from eating affected fish. (See Part 2, Ciguatera Fish Poisoning.)

Prevention

Barracuda usually keep their distance from swimmers and divers. Some people theorize that barracuda attacks occur when the predator mistakes a human for prey. Because the flash of jewelry or camera equipment may be viewed by the barracuda as a silvery fish, removing jewelry and avoiding murky water may reduce the chance of an attack.

Signs and Symptoms

The sharp teeth of a barracuda can cause straight or V-shaped cuts, with resulting tissue loss, bleeding, and possible wound infection.

Numbness or inability to move a finger or toe normally is often a sign of tendon or nerve damage that requires repair.

 # First Aid

For minor bites, gently pull the edges of the skin open and remove any embedded teeth either by rinsing or using tweezers. Then scrub directly inside the wound with gauze or a clean cloth soaked in clean, fresh water. Press on the wound to stop bleeding. If bleeding persists or if the edges of a wound are jagged or gaping, the victim probably needs stitches. Taping a bite shut is often an effective alternative but may leave a more visible scar than suturing. For more information about wound care, see Part 2, *Staph, Strep,* and General Wound Care.

Victims who appear pale, sweaty, and nauseated are in danger of fainting. Lower the victim to the ground.

Barracuda can sever arteries or veins. In such cases, victims can die rapidly from blood loss. Often, a rescuer can stop bleeding from severed blood vessels by applying firm pressure directly on the wound with anything handy (swimsuit, towel, hand). Pressure usually causes the vessel to clamp down in spasm, and clots begin to form. In the water, this procedure can be nearly impossible, especially while helping a victim to shore or to a boat. In these cases, when bleeding might be fatal, a tourniquet is appropriate. Tying a surfboard leash or dive mask strap around a massively bleeding limb could save a person's life.

Help a bleeding victim get out of the water as quickly as possible. At the beach or in the boat, control the bleeding by pressing directly on the wound, then remove any tourniquets. (Leaving a tourniquet on can cause permanent damage.) Maintaining pressure on the wound, take the victim to an emergency room as quickly as possible.

Advanced Medical Treatment

Barracuda teeth often break off inside wounds and are visible on x-ray. Remove the teeth meticulously to prevent infection and foreign body granuloma. Examine for tendon and nerve damage and repair, or refer, as necessary.

Thoroughly scrub, explore, irrigate, and debride all barracuda bites. Suturing bites to control bleeding, preserve function, or improve appearance is appropriate.

Do not prescribe antibiotics for minor bites with no sign of infection, except in immune-compromised patients. Even though using antibiotics to decrease the incidence of marine wound infection is unproven, early antibiotic therapy is reasonable for patients with large, deep lacerations. No clinical-outcome data favor a particular antibiotic regimen for decreasing risk of marine infections. For more information about antibiotic therapy, see Part 2, *Staph, Strep,* and General Wound Care.

For trauma victims in hemorrhagic shock, use the airway, breathing, and circulation (ABCs) protocol of trauma management.

BILLFISH *(A'U)* WOUNDS

Billfish, or *a'u,* are general names for marlin, sailfish, spearfish (family Istiophoridae), and swordfish (family Xiphiidae). Each of these fish has an extended, pointed upper jaw that is longer than the lower. The swordfish has the longest bill compared to body length. Unlike most billfish,

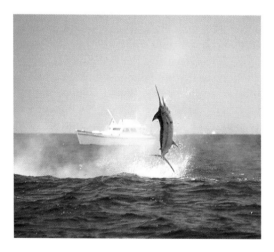

One of the reasons anglers prize billfish is because they fight for their lives so fiercely. During the struggle, the powerful fish are capable of inflicting serious, sometimes fatal, wounds on their captors. Kona Coast. (Richard Brill)

While the anglers gaffed it, one of these two marlins punctured the keel of the inflatable dinghy. Kailua Bay, O'ahu. (Tony Barcia and Peter Barcia, M.D.)

which have round bills, the swordfish bill is flat, with sharp edges, and is about one-third the fish's total body length. In spearfish, the upper jaw is considerably shorter than in other family members, extending only a short distance beyond the lower jaw.

All billfish are fast-swimming carnivores that speed through schools of tuna and squid, slashing the bill back and forth as they go. This process stuns or injures the prey, which the billfish then circles back to eat.

Some billfish skewer their prey. Swordfish have occasionally rammed ships and boats with their swords, usually killing themselves in the process.

Hawai'i's waters host six species of billfish, ranging from 50 to nearly 2,000 pounds. All are highly prized gamefish. Researchers believe the waters off the Kona Coast of the island of Hawai'i are spawning sites for some billfish species.

Mechanism of Injury

In the water, billfish use their bills to slash and stun prey. When caught by a hook, a billfish fights back with a similar technique, flailing its head back and forth. As the angler hauls the fish alongside a boat or on deck, this slashing motion sometimes punctures and cuts its human foe. A seemingly exhausted or dead fish occasionally will "come alive" in the cockpit of a boat, surprising and injuring someone nearby.

Billfish injuries in Hawai'i have ranged from "bill-whacking" cuts and bruises to fatal punctures.

Incidence

Recreational trollers are at the highest risk for billfish injuries, which occur during the landing process. Each year, some victims speared in the hand or leg undoubtedly go unreported.

Reported injuries are often dramatic. In 1995, a live billfish punctured the thigh of a sport fisherman still strapped to his chair. Several years earlier, a marlin speared a Honolulu sport fisherman in the lung. The man had the fish close to the boat, ready to land, when it rose up, striking him in the chest. The man survived. In the mid-1980s, a billfish stabbed a Kona fisherman under the chin, piercing his jaw and right eye. He lost the eye but lived. In the early 1980s, researchers on Hawai'i's remote Kure Atoll found a Boston Whaler washed ashore with a marlin spike through the hull and into one of the seats. The operator was not found.

Billfish caught by commercial long-liner fishing boats are usually dead when workers pull the fish aboard. One fisherman, however, died in Hawai'i in 1996 when a fighting swordfish punctured his right eye with its bill.

Prevention

Never assume a freshly landed billfish is harmless, even if it appears exhausted or dead. If a fish comes to the boat fighting strongly, many anglers loosen the drag and let the fish fight a little longer. This tires the fish before its landing. Others use a weapon such as a bat to stun or kill the fish before hauling it aboard. Some anglers saw off a landed fish's bill as soon as possible.

Signs and Symptoms

Scrapes and cuts from billfish may bleed. Redness and swelling signal an infection.

A puncture wound can penetrate vital organs, causing internal bleeding or other potentially fatal wounds. Seriously injured victims may become pale, sweaty, confused, and short of breath.

 # First Aid

For minor wounds, gently pull the edges of the skin open and remove any embedded material either by rinsing or using tweezers. Then scrub directly inside the cut with gauze or a clean cloth soaked in clean, fresh water. Press on the wound to stop bleeding. If bleeding persists or if the edges of a wound are jagged or gaping, the victim probably needs stitches. Taping a cut shut is often effective but may leave a more visible scar than suturing. For more information about wound care, see Part 2, *Staph, Strep,* and General Wound Care.

Victims who appear pale, sweaty, and nauseated are in danger of fainting. Lower the victim to the deck.

For a billfish puncture, never remove the bill from the wound. If the bill has penetrated a major blood vessel or vital organ, leaving the bill in place may save the victim's life. Cut or saw the bill off the fish, then rush the person to an emergency room. If bleeding occurs around the bill, press gently but steadily against the opening with any clean cloth.

If the struggling fish pulls its bill out of the wound, hold any clean cloth firmly against the opening while rushing the victim to an emergency room.

 ## Advanced Medical Treatment

Billfish bills can break off inside wounds and are visible on soft-tissue x-ray. Ultrasound may also help locate pieces of bill. Remove them meticulously to prevent infection and foreign body granuloma.

Thoroughly scrub, explore, irrigate, and debride all wounds. Examine for tendon and nerve damage and repair, or refer, as necessary. Suturing wounds to control bleeding, preserve function, or improve appearance is appropriate.

Do not prescribe antibiotics for minor injuries with no sign of infection, except in immune-compromised patients. Even though using antibiotics to decrease the incidence of marine wound infection is unproven, early antibiotic therapy is reasonable for patients with large, deep wounds. No clinical-outcome data favor a particular antibiotic regimen for decreasing risk of marine infections. For more information about antibiotic therapy, see Part 2, *Staph, Strep,* and General Wound Care.

Have a surgeon in attendance during removal of bills embedded in vital structures. For victims in hemorrhagic shock, use the airway, breathing, and circulation (ABCs) protocol of trauma management.

BOX JELLYFISH STINGS

Box jellyfish belong to a jellyfish group called Cubomedusae, found only in tropical and subtropical waters. Cubomedusae have transparent, almost perfectly square bells. Their bottom edges are straight rather than scalloped, and tentacles hang from each of the bell's four corners.

In the middle of each of the four, flat sides of a box jellyfish bell lie the animal's sensory organs, including elaborate, well-developed eyes. In darkness, a box jellyfish can detect light from a match 4 or 5 feet away and will swim toward it. These creatures are sensitive to bright sunlight, though, apparently retreating to deep water at midday.

Box jellyfish *(Carybdea alata)* usually appear along Hawai'i's leeward shores eight to ten days after the full moon. (Waikiki Aquarium photo)

Box jellyfish can swim at least 2 miles an hour. They feed mainly on shrimp and fish, catching and stinging them with cells called nematocysts, abundant on the trailing tentacles.

Box jellyfish are infamous in Australia and surrounding areas for their sometimes fatal stings. The most dangerous species found there, the sea wasp *(Chironex fleckeri),* does not inhabit Hawaiian waters. Three kinds of stinging but nonlethal box jellyfish appear periodically in Hawai'i's bays and along the shorelines. Two are *Carybdea alata,* about 3 inches tall and 2 inches wide, and *Carybdea rastoni,* about 1 inch square. Both have pinkish tentacles trailing from the four corners of the square. The tentacles of the larger species can be up to 2 feet long. Researchers are working to identify the third species.

Mechanism of Injury

An undischarged nematocyst looks like a narrow tube folded inside a capsule. When the "capsule" receives a touch, or a chemical signal, lid-like flaps covering one end burst open, and the nematocyst ejects like a jack-in-the-box.

The authors found this box jelly-fish *(Carybdea alata)* swimming in the Ala Wai Boat Harbor eight days after the full moon. The body of the jellyfish is safe to touch; the four trailing tentacles, which bear stinging cells, are not. (Craig Thomas, M.D., and Susan Scott)

A discharged nematocyst looks like a bulb bearing a hollow thread. The thread penetrates the skin of prey or humans, after which venom oozes from the bulb. Surrounding blood vessels pick up the venom and carry it through the circulatory system.

Nematocyst venom contains an impressive number of active compounds, both protein and nonprotein. Chemical transmitters similar to the ones that trigger nerve conduction, cardiac contraction, and allergic reactions are among these compounds.

No one is immune to nematocyst venom. Some people are allergic. Delayed skin allergies are common.

Incidence

Honolulu city and county lifeguards have recorded sting incidents on Oʻahu beaches. Although the records do not identify which species caused a sting, lifeguards estimate that approximately 90 percent of the stings they treat at the beaches of Waikīkī, Ala Moana, and the Waiʻanae Coast come from box jellyfish. Estimates of Hanauma Bay stings are about 50 percent box jellyfish and 50 percent Portuguese man-of-war. Windward stings are nearly 100 percent Portuguese man-of-war. In 1994, lifeguards treated approximately eight hundred victims of box jellyfish stings. Twenty to thirty were major stings, requiring additional medical treatment.

Box Jellyfish Stings on O'ahu, 1994

Beach	Minor Stings	Major Stings
Ala Moana	112	5
Waikīkī	382	6
Hanauma Bay	86	12
Māʻili	30	2
Nānākuli	10	1
Pōkaʻi	120	1
Mākaha	25	1

Hawai'i's box jellyfish usually follow a monthly cycle. They show up in leeward waters eight to ten days after the full moon. They stay for about three days, then disappear until the next month. No one knows exactly why, but the pattern is so regular that lifeguards often correctly predict their appearance. Stings seem to occur more frequently in the morning.

During box jellyfish invasions, dozens of swimmers get stung in all ranges of severity. Most of these stings are treated at the beach, with no complications.

Prevention

Lifeguards usually post signs on beaches when box jellyfish are in an area. Heed the warnings, and stay out of the water. Do not count on simply avoiding the jellyfish while swimming. These transparent creatures are difficult to spot in the water. Beached, they look like the cellophane from a cigarette pack.

If you are not sure whether jellyfish are in an area, ask lifeguards. If you do swim in jellyfish-posted waters, or in areas where they often appear, wear as much protective clothing as possible. Double pantyhose or Lycra body suits help prevent stings; coating the skin with petroleum jelly (or any other substance) does not.

Signs and Symptoms

Symptoms of box jellyfish stings range widely from person to person. The severity of a sting depends on how much skin was exposed and for how long, and the size and sensitivity of the victim. Because children and small adults receive more venom per pound of body weight than larger people, they tend to have more intense reactions. The body part injured can also determine a person's reaction. Eyes and facial areas are usually more sensitive than arms and legs. Depending on these factors, box jellyfish stings may range from no visible marks on the skin to slight redness, hivelike welts, and blisters.

The classic box jellyfish sting in Hawai'i causes immediate, sometimes severe, pain and a burning sensation lasting ten minutes to eight hours. At the same time, a red hivelike welt appears at the sting site. Sometimes, stings appear as raised red whips of fine, parallel lines across the affected skin. Box jellyfish marks typically last from two hours to three days. With extensive stings, some victims suffer chills, fever, malaise, vomiting, and abdominal pain. Sometimes, lymph nodes in the vicinity of the sting swell. This is usually a sign of the toxic effect of the venom, or it may be an early sign of infection. Rarely, a sting will cause nerve damage, resulting in numbness or weakness of the affected area for weeks to months. A scar or permanent discoloration of the skin occasionally appears at the site. Eye stings can cause pain, irritation, swelling, tearing, blurred vision, or light sensitivity.

Some people have severe reactions to box jellyfish venom. These responses are usually immediate and can be life threatening. Signs of this reaction are irregular heartbeat, difficulty breathing, mental disturbances, high or low blood pressure, and weakness.

Delayed skin allergies after box jellyfish stings are common, some persisting for months. In a study of stings of Hawai'i's smaller box jelly-

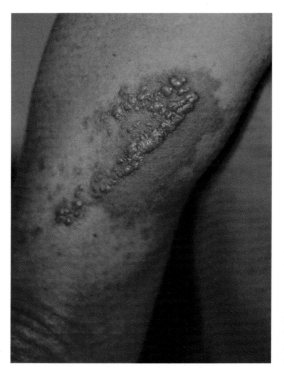

The persistent rash of this box jellyfish sting on the arm of a Honolulu man lasted more than ten months. The red area around the blisters is an allergic reaction to the prescription medicine doxepin, which was applied to the sting. (Allan Izumi, M.D.)

Some jellyfish nemato-
cysts are too weak to
sting most humans. Both
of us have often handled
these moon jellies
(*Aurelia* species) in the
Ala Wai Boat Harbor in
Waikīkī without suffer-
ing pain or ill effects.
Children, or particularly
sensitive adults, how-
ever, may feel a sting.
(Susan Scott)

fish species *(Carybdea rastoni),* more than half the subjects suffered skin
rashes that appeared one to four weeks after the initial sting. These reactions cause itching, but do not hurt.[1] Stings from box jellyfish are usually
more painful and longer lasting than from Portuguese man-of-war.

✚ First Aid

Two concepts are key to treating box jellyfish stings. One is to prevent
the firing of any undischarged nematocysts remaining on the skin,
which will prevent the injury from getting worse. The second is to treat
the symptoms and pain caused by already fired nematocysts.

The following first aid treatment, based on current Australian
research, is recommended for the stings of all species of box jellyfish:

1. Immediately flood the area with household vinegar to keep
   undischarged nematocysts from firing. This does not relieve
   pain but prevents additional stings.[2] Never rub the area with
   sand or anything else.

2. Irrigate eye stings with copious amounts of room temperature
   tap water for at least fifteen minutes. Do not use vinegar in eyes.

If vision blurs or the eyes continue to tear, hurt, and swell or are light sensitive after irrigating, see a doctor.

3. Pluck off any vinegar-soaked tentacles with a stick or other tool.

4. If the victim has shortness of breath, weakness, palpitations, muscle cramps, or any other generalized symptoms, take him or her to an emergency room.

5. For pain relief, apply ice packs.[3] If pain becomes unbearable, take the victim to an emergency room.

No studies support applying heat to box jellyfish stings. Contradictory studies exist on the effectiveness of meat tenderizer, baking soda, papaya, or commercial sprays (containing aluminum sulfate and detergents) on nematocyst stings. These substances may cause further damage. Some kinds of meat tenderizer, for example, can cause skin peeling. In one U.S. fatality from the box jellyfish *Chiropsalmus quadrumanus,* rescuers almost immediately applied meat tenderizer to the affected arm. The child was soon comatose and later died. Alcohol and human urine are common nematocyst remedies but can be harmful. An Australian study reports that alcohol and urine caused massive discharge of box jellyfish nematocysts.[4]

Most Hawai'i box jellyfish stings disappear by themselves. Because of this, even harmful therapies can appear to work. A key concept in the first aid of any injury is: Do no harm. Therefore, avoid applying unproven, possibly harmful substances to stings.

Nematocyst toxins occasionally cause lymph nodes near a sting site to swell. If a red streak develops between the two areas, or if either area becomes red, warm, and tender, see a doctor immediately. For more information about infection, see Part 2, *Staph, Strep,* and General Wound Care.

Few box jellyfish stings in Hawai'i cause life-threatening reactions, but it is always a possibility. Some people are extremely sensitive to the venom; a few have allergic reactions. Regard even the slightest breathing difficulty or altered level of consciousness as a medical emergency. Call for help and use an automatic epinephrine injector if available. (See Part 2, Allergic Reactions.)

Advanced Medical Treatment

The specific treatment for inactivating undischarged box jellyfish nematocysts is vinegar dousing. Apply ice for pain relief. No clinically useful diagnostic tests exist for Hawai'i's box jellyfish stings. Australian box jellyfish (sea wasp) antivenom is not appropriate for our species and is not available in Hawai'i. Hawai'i has no reported fatalities from box jellyfish stings.

Rarely, the skin blisters. Treat these patients for partial-thickness burns. Recurrent dermatitis with an itching rash but no pain is common in box jellyfish stings. Therapy is unproven. For persistent cases, try oral steroids for several days.

A few people have anaphylactic reactions to nematocyst stings. These patients require standard therapy of epinephrine, antihistamines, steroids, rehydration, and airway support. For more information, see Part 2, Allergic Reactions.

No literature exists on box jellyfish stings to the eye. Eye stings from the sea nettle, a jellyfish common in Chesapeake Bay, are usually self-limited, disappearing in twenty-four to forty-eight hours. One report recommends using topical steroids and cycloplegic drops. Some victims have persistent dilated pupils with blurred vision, particularly for nearby objects. Initially, intraocular pressure is elevated. This pressure remains high in some patients, producing glaucoma in the affected eye.[5]

A severe reaction to Australia's box jellyfish venom is cardiac toxicity presenting as arrhythmias, hypotension, and conduction delay. These symptoms may be caused by the toxin's effect on ion channels. Many of these victims respond to IV fluids. Calcium channel blockers during resuscitation, however, may be beneficial. Calcium may increase cardiac toxicity.[6]

Several cases of peripheral nerve deficits (motor and sensory) after jellyfish stings have been reported. They all developed several days after the sting, affected nerves near the sting site, were not associated with impaired circulation, and improved over several months without treatment.[7]

CONE SNAIL STINGS

Of the three hundred types of cone snails (*Conus* species) in the world, 91 percent are found in the Pacific Ocean, 69 percent in the Indian Ocean, and 15 percent in the Atlantic Ocean. These beautiful snails are most abundant in the warm waters of the tropics and subtropics. Hawai'i hosts approximately twenty-five species, ranging from the shallow reefs to deep, offshore waters. Cone snails are from 3/4 inch to several inches long, depending on age and species.

These slow-moving snails do not conjure up an image of effective hunters, but they are. All cone species are predators, falling into one of three groups: worm hunters, snail hunters, and fish hunters. The snails ambush unsuspecting prey, then inject a powerful neurotoxin that quickly paralyzes their victim.

The textile cone shell *(Conus textile)* is the most toxic of Hawai'i's cone shells. It is best not to touch these snails; if you do, always hold them at the wide end. Never put a cone shell into a pocket in swim trunks or a dive jacket. Hanauma Bay. (D. R. and T. L. Schrichte)

Cone snails are nocturnal, actively hunting at night and lying hidden in sand or under rocks during the day. In arming itself, the snail drops a hollow tooth, resembling a barbed harpoon, from a tooth-storage sac into its throat. The creature's venom duct fills the tooth with venom, and the loaded tooth then moves into the snail's proboscis, a projectable, flexible tube resembling a tiny elephant trunk.

When a potential prey comes within reach, the cone snail points with the proboscis and jabs the venom-filled tooth into its victim like a harpoon. In snail hunters, the tooth comes free of the cone snail. In the fish and worm hunters, the tooth remains connected to the snail. After stinging its prey, the cone snail expands its mouth around the animal and digests it.

The cone snails most dangerous to humans are the fish hunters, followed by the snail hunters, then the worm hunters. Many of Hawai'i's common cone snails are worm hunters, the least dangerous.

Mechanism of Injury

Cone snail toxicity varies depending on species and size. More than one hundred conotoxins have been identified so far. Most are peptides (the building blocks of proteins) that work together to block muscle contractions, nerve impulses, and communication between nerve and muscle. These peptides also break down proteins.

The hollow, transparent tooth of a cone snail can remain lodged in the puncture wound. In a 2-inch-long textile cone snail, the tooth is about 1/4 inch long.

Incidence

Of the three hundred species of cone snails in the world, eighteen have stung humans. At least eight species are found in Hawai'i, and the common, beautiful textile cone, *Conus textile,* is the most dangerous. This snail-eating species has caused at least two, and possibly more, fatalities in other parts of the world. No fatalities have been reported in Hawai'i from this or any other cone snail species, but stings do occur occasionally. Approximately thirty fatalities have been recorded worldwide from all cone snail species combined.

Prevention

To be safe, regard all cone snail species as dangerous. The best approach is not to pick up cone snails at all, but if you must, never hold them on or near the narrow end. Even holding the wide end, your bare hand is not always safe; the mouth tube of some cone snails is flexible enough to reach around and jab it.

Never put a cone snail into a pocket or assume you are safe with gloves on. These creatures' teeth can penetrate fabric.

If you want to look at a cone snail that is partially buried in the sand, roll it out of the sand with a tool rather than using your bare hand. Then put it back where you found it.

Signs and Symptoms

Symptoms vary greatly depending on the type and size of the snail. The initial pain from a worm-hunting cone snail puncture usually is similar to a bee or wasp sting. With no treatment, this disappears in about two hours. The puncture wound may be tender to the touch but is otherwise unremarkable. Numbness of the hand or finger almost always follows the burning sensation of the initial sting. Some victims do not feel a sting at all but still experience numbness, sometimes for weeks.

Cases of envenomation by textile cone snails (and sometimes other cone snails) may be more severe. Reactions can include shortness of breath, headache, stomach cramps, and nausea for up to twelve hours. One eight-year-old New Guinea patient developed gradual paralysis two hours after a textile cone snail envenomation. This resulted in slurred speech, shallow breathing, absence of reflexes, and a fast heart rate. The child eventually stopped breathing. Doctors inserted a tube in her trachea to support breathing. In two hours, she regained consciousness. In twenty hours, she had fully recovered.

If a cone snail wound becomes infected, the tooth may still be in the wound.

 # First Aid

Most cone snail punctures in Hawai'i are not dangerous and require no specific treatment other than thorough scrubbing of the wound. Rarely, a person stung by a cone snail experiences nausea, headache, or difficulty breathing. If any of these symptoms develop after a sting, take the victim directly to an emergency room. En route, position the bite site below the rest of the body while keeping the victim as still as possible. Apply a broad pressure bandage over the bite about as tight as an elastic wrap to a sprained ankle. This slows the spreading of venom through the lymph system.[1] Monitor fingers or toes for pink color and warmth to be sure arterial circulation is not cut off by too much pressure.

If possible, take the stinging snail to the emergency room, being careful it does not sting again.

 # Advanced Medical Treatment

No antivenom exists for cone snail envenomations, and no clinically useful tests exist for diagnosis. Remove occlusive dressing. If needed, use buffered bupivacaine without epinephrine for pain control. Some doctors believe hot water soaks to tolerance help relieve pain. In one study, however, heating cone snail venom to boiling did not inactivate the toxins.[2]

Explore the wound for a retained tooth. Soft-tissue x-ray or ultrasound may help locate an occult tooth. For information about wound care, see Part 2, *Staph, Strep,* and General Wound Care.

Serious reactions to Hawai'i's cone snails are rare. Some cone snail toxins target a diverse set of ion channels and neuronal receptors, including some calcium channels.[3] Others contain the neurotransmitter serotonin.[4]

Ascending paralysis resulting in respiratory distress has been reported.[5] Prepare for airway management, including possible intubation, for respiratory failure. Paralysis resolves spontaneously. No proven drug therapy exists.

Save the cone snail for identification by the Hawaii Department of Health.

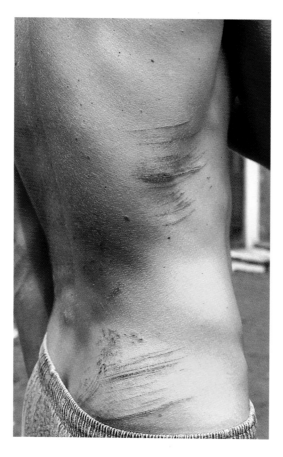

One-day-old coral scrapes on the back of a Hale'iwa surfer show the raised, red welts common in coral abrasions. (Craig Thomas, M.D., and Susan Scott).

CORAL CUTS AND STINGS

Each single coral animal has a soft, hollow body shaped like a vase. The bottom of the vase is attached to the ocean floor or to the stony skeletons of other corals. In most species, the animal's mouth, at the top, opens to the outside and is surrounded by tentacles, which bear stinging cells called nematocysts that paralyze and capture passing food.

Stony corals build a calcium carbonate (limestone) skeleton, secreted by the lower skin layer of the soft body. This process creates a hard cup with the soft body attached inside. Most corals form colonies that expand by budding new bodies. All individuals of a colony are connected to one another by a thin sheet of tissue covering the entire surface of the coral colony.

Many stony corals are carnivores, capturing planktonic life with their tentacles. Other stony corals collect drifting animal particles in mucous films or strands covering the colony. Tiny, beating hairs move trapped food particles to the mouths.

Almost all reef-building corals house algae, which give corals their color. Stony corals and their algae have a symbiotic relationship: The algae get carbon dioxide and other waste products from the breakdown of the corals' prey; the corals get glucose and other nutrients generated by the algae.

Although people in Hawai'i sometimes believe they have been stung by fire coral *(Millipora),* these coral-like species have not been reported in Hawaiian waters.

Mechanism of Injury

The limestone skeletons of many corals are sharp, sometimes cutting and scraping skin with only the slightest contact. Small cuts and abrasions from corals are often slow to heal and prone to infection. This is caused by a combination of factors:

♦ Tiny pieces of coral skeleton and other debris often lodge inside cuts.

♦ Toxin from the animals' tentacles delays healing.

♦ Bacteria-laden seawater bathes the wound.

♦ People who get coral cuts often return to the water before healing occurs.

Nerve, tendon, and bone damage occasionally occurs in deep coral cuts.

Contrary to popular belief, coral cannot grow inside a coral cut. Bacteria, however, can and often do. This, combined with pieces of the coral's soft body, skeleton, mucus, and algae, often cause coral cuts to become infected and heal slowly.

Because most corals use stinging cells (nematocysts) to catch food, they can also sting people. Depending on the type of coral and a person's individual reaction, stinging cells from coral tentacles can produce red, hivelike welts at the slightest touch. The reactions may often be delayed, worsening overnight.

Incidence

Coral cuts, infections, and skin reactions are extremely common in Hawai'i, where reefs surround the islands.

Prevention

Wet suits prevent accidental skin contact with coral, and gloves provide protection from coral cuts. It is better for the health of the coral, however, to avoid touching coral heads when snorkeling or diving.

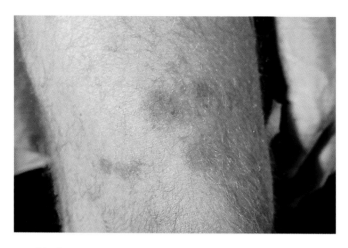

Hivelike bumps are typical of a "coral kiss." A brush with coral while windsurfing in Kailua Bay did not break the leg's skin, but the welts lasted nearly two weeks. (Susan Scott)

When surfing, bodyboarding, or windsurfing, keep your feet off the reef. Wearing reef shoes helps, but remember that standing on coral can cause irreversible damage to the colonies.

Signs and Symptoms

Coral cuts can be any size or shape. Sometimes, a coral laceration is redder and more tender than the injury warrants because the skin reacts to the coral's stinging cells. Severe reactions include welts and blisters at the site of contact. Red, raised welts may be symptoms of a reaction to the coral's nematocysts or may be signs of an early infection.

Even with prompt treatment, coral wounds may become infected, have skin breakdown, and heal slowly. Coral cuts can take three to six weeks to heal, and often leave scars.

 # First Aid

For minor cuts, gently pull the edges of the skin open and remove embedded coral either by rinsing or using tweezers. Then scrub directly inside the cut with gauze or a clean cloth soaked in clean, fresh water. Press on the wound to stop bleeding. If bleeding persists or if the edges of a wound are jagged or gaping, the victim probably needs stitches. Taping a cut shut is often an effective alternative but may leave a more visible scar than suturing. For more information about wound care, see Part 2, *Staph, Strep,* and General Wound Care.

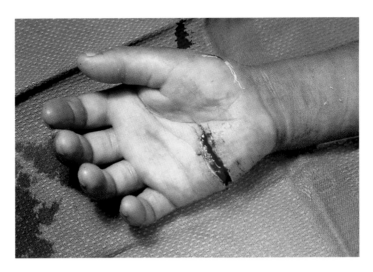

This patient cut his hand on an unidentified species of stony coral while snorkeling off O'ahu. The accident lacerated the ulnar nerve. (K. Reicker, M.D.)

It is not true that using povidone-iodine or other iodine solutions to wash coral cuts will cause coral to grow in the wound. Coral is a marine animal; it can never grow inside the human body.

A popular Island treatment for coral cuts is to urinate on the wound. This sometimes washes broken pieces of coral from the cut, but urine discharges nematocysts in Australia's box jellyfish.[1] Urine may therefore make the injury worse.

For large cuts, fever, or any other sign of infection or illness after coral contact, see a doctor.

 # Advanced Medical Treatment

Localized reactions are almost always the primary concern in coral cuts and stings. Thoroughly debride the area and explore for pieces of coral. Consider soft-tissue x-ray or ultrasound for deep cuts. Superficial infections can develop days after the injury. Topical antibiotics such as mupirocin may decrease the incidence of infection. Reserve systemic antibiotics for patients with compromised immune systems, signs of infections, or large wounds.

Fever after a coral cut usually indicates infection. These patients need wound exploration and debridement. Use cultures and antibiotics as described in Part 2, *Staph, Strep,* and General Wound Care. These patients are also at risk for *Vibrio* and *Mycobacterium marinum* infections.

CRAB PINCHES

All crabs have five pairs of legs. In most species, the front two legs are heavier than the others, ending in a claw or pincer. With the pincers, crabs pass food to the feeding appendages, above and beside their mouth. The appendages tear the food into smaller pieces and direct them into the mouth.

The shape of a crab's legs and pincers often reflect its feeding habits. Hawai'i's box crabs, or *pokipoki* (*Calappa* species), have one crushing claw and one cutting claw used for opening and eating hard-shelled invertebrates. These crabs average 3 inches across. The "7-11" crab, or *alakuma (Carpilius maculatus),* and other members of this family, also have large pincers for breaking open shelled animals.

Although other invertebrates, such as shrimp and snails, are common prey for many crabs, most crabs combine hunting with scavenging for dead animal material.

Many crabs are restricted to walking and running, but some can swim. In swimming crabs, the last pair of legs are paddle-shaped. Swimming crabs, such as the Samoan crab *(Scylla serrata),* Hawaiian crab *(Podophthalmus vigil),* and white, or *kuahonu*, crab *(Portunus sanguinolentus),* are popular food sources in Hawai'i. These adult crabs range from about 4 to about 7 inches across the back shell. Several have powerful pincers.

Most crabs flee when a human approaches. But if cornered or captured, crabs will fight a brave battle, using their pincers as weapons.

The swimming crab, or *'ala'eke (Charybdis hawaiiensis),* has claws adapted for grasping food. If a person corners the crab, the claws may also grasp human skin. Hanauma Bay. (D. R. and T. L. Schrichte)

Hawaiʻi's ghost crabs, or *ʻōhiki (Ocypode ceratophthalmus),* use their claws primarily for carrying food and digging holes in the sand. When threatened, they raise the claws in defense. Hanauma Bay. (D. R. and T. L. Schrichte)

Mechanism of Injury

Hawaiʻi's crabs contain no venom, but the claws can bruise, crush, puncture, or cut. The larger the pincer, the more extensive the wound.

Some people in the tropical Pacific have become seriously ill after eating crab species closely related to Hawaiʻi's "7-11" crabs. No Hawaiʻi species has yet been shown to be toxic.

Incidence

Crabbing is a minor fishery in Hawaiʻi; injuries from these creatures are infrequent.

Prevention

Wear heavy gloves when handling crabs.

Signs and Symptoms

Most crab injuries look like a small slash wound or a bruise, often on the hand.

 # First Aid

For minor cuts, gently pull the edges of the skin open and scrub directly inside the cut with gauze or a clean cloth soaked in clean, fresh water. Press on the wound to stop bleeding. If bleeding persists or if the edges of a wound are jagged or gaping, the victim probably needs stitches. Taping a cut shut is often effective but may leave a more visible scar than suturing. For more information about wound care, see Part 2, *Staph, Strep,* and General Wound Care.

If a finger or toe is numb or will not move normally after a bite, see a doctor immediately.

Advanced Medical Treatment

Injuries from crabs are generally minor.

Thoroughly scrub, explore, irrigate, and debride all lacerations. Do not prescribe antibiotics for minor injuries with no sign of infection, except in immune-compromised patients.

Suturing wounds to control bleeding, preserve function, or improve appearance is appropriate. For more information, see Part 2, Advanced Medical Treatment in *Staph, Strep* and General Wound Care.

CROWN-OF-THORNS SEA STAR STINGS

Crown-of-thorns sea stars *(Acanthaster planci)* have seven to twenty-three arms. Although these sea stars (also called starfish) can grow to 24 inches across, the average size is around 12 inches.

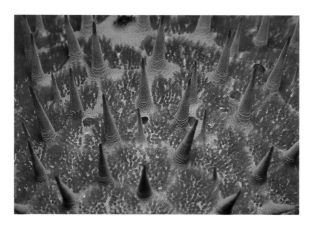

Crown-of-thorns sea star spines. The skin covering the spines secretes mucus that contains venom. In puncture wounds, some of this venomous mucus enters the skin, causing a burning sensation. Such injuries are rarely serious, disappearing on their own in twenty-four to forty-eight hours. Hanauma Bay. (D. R. and T. L. Schrichte)

Crown-of-thorns sea stars eat the soft bodies of coral. This sea star at Hanauma Bay has wrapped itself around a projection of finger coral *(Porites compressa)*. (D. R. and T. L. Schrichte)

On the lower surface of these nocturnal animals lie rows of yellow, suction-tipped tube feet, harmless to human skin. The top surface of this sea star is covered with brittle, sharp spines averaging about an inch long. It is these venom-covered spines that can cause painful puncture wounds. Crown-of-thorns sea stars are slow-moving, nonaggressive animals; they use their venom only for defense.

This sea star eats coral. At night, the greenish-to-reddish creature uses its tube feet to walk to a coral head. The sea star everts its stomach directly onto the coral, digesting the soft tissue from the coral's cuplike skeleton. In one night, a crown-of-thorns can eat an area the size of its own body, leaving behind the coral's dead, white framework.

Mechanism of Injury

The skin covering the top of the sea star, including the spines, secretes mucus that contains a venom with multiple toxins. When a spine penetrates human skin, some of this venomous mucus is left in the wound.

Each spine of these sea stars has three raised cutting edges at the tip. Crown-of-thorn punctures are frequently on the hands, forearms, feet, and lower legs.

Incidence

Crown-of-thorns sea stars normally number about three to five per square mile throughout their range in the tropical and subtropical Indian and Pacific Oceans. In some areas, these sea stars occasionally reach abnormally high numbers, up to fifteen per square yard, capable of devastating large areas of reef. Some people speculate that the population explosions are caused by human alteration of the environment, but this remains in question.

Hawai'i's reefs have not suffered huge, destructive blooms of crown-of-thorns sea stars, but they sometimes live in higher concentrations than usual in certain areas of the islands.

Because most people with these wounds do not seek medical attention, it is difficult to gauge their frequency.

Prevention

If a person is extremely careful, it is possible to handle this sea star with bare hands. But the slightest pressure against a spine will puncture human skin.

Wearing gloves is the safest way to touch crown-of-thorns, but remember that the spines are sharp enough to poke through neoprene, leather, and even the soles of sneakers. Windsurfers, wave riders, swimmers, and divers should keep their hands and feet off the reef. Crown-of-thorns are sometimes difficult to see. Their colors can blend well with their backgrounds.

Signs and Symptoms

Within minutes, a crown-of-thorns puncture causes an acute burning sensation. A bluish color soon appears 1/4 to 1/2 inch around the puncture site. The wound often bleeds more than the injury warrants.

In four to six hours, the area becomes red and swollen, sometimes an inch or more around the wound, and the pain lessens to a dull ache. Acute pain disappears several hours after the injury, but the puncture site remains tender. After twenty-four hours, most victims feel a numbness an inch or more around the puncture site, which is usually still red and swollen. The symptoms gradually go away without treatment. After forty-eight hours, it is often difficult to locate the wound.

Some victims may be extremely sensitive to crown-of-thorns toxin, and the possibility of an allergic reaction always exists. A general feeling of illness or any difficulty breathing are signals of a potential allergy to the toxin. Multiple punctures with embedded spines may result in acute reactions, including numbness, tingling, weakness, nausea, vomiting, swollen lymph nodes, and paralysis.

Crown-of-thorns sea star spines are brittle and can break easily. If a spine snaps near its base, the skin of the sea star is often strong enough to allow the spine to remain attached to the animal. If possible, withdraw a punctured hand or foot slowly to prevent the spine from staying embedded in the wound.

First Aid

Remove any obvious broken spines protruding from the wound.

Unless a victim is allergic to the venom, these wounds are not a medical emergency and can be treated at home. Soaking the wound in hot water for pain control is unproven. Some doctors believe soaking the sting site in hot, nonscalding water for thirty to ninety minutes may reduce pain caused by the venom.[1] Others believe heat application has no benefit in crown-of-thorns injuries and should not be used.[2] For more information, see Part 2, First Aid in *Staph, Strep,* and General Wound Care. Infections from these wounds are rare.

If you cannot remove broken spines or if you develop hives, breathing difficulty, numbness, or weakness, go to an emergency room. The symptoms usually disappear after the spines are removed.

Advanced Medical Treatment

No specific antidote or diagnostic tests exist for crown-of-thorns envenomation.

These puncture wounds are generally not life threatening, even with multiple stings. The primary issues in most incidents are pain control, spine removal, and wound debridement. Use buffered bupivacaine to manage pain. Soft-tissue x-ray or ultrasound may detect embedded spines. Surgical exploration may be necessary. Reserve antibiotics for immune-compromised patients or those with signs of infection. For more information, see Part 2, *Staph, Strep,* and General Wound Care.

A rare complication is prostaglandin-mediated hypotension. Large amounts of crown-of-thorns venom decreases systemic vascular resistance and cardiac output. Indomethacin blocks this venom-induced hypotension. In animal studies, venom injection causes thrombocytopenia, leukopenia, and hepatitis. The clinical significance of this is unclear.[3] In humans, systemic symptoms disappear with removal of the spines.[4]

FIREWORM STINGS

Fireworms, sometimes called bristleworms, belong to a large group of marine worms, Polychaeta, which are segmented like earthworms. In the polychaetes, however, each body segment bears a pair of tiny paddles, or "feet," protruding from either side. Bundles of bristles extend

A Hawai'i fireworm. These worms get their name from the burning sensation their bristles create after puncturing human skin. Hanauma Bay. (D. R. and T. L. Schrichte)

from the end of each foot like little whisk brooms. The scientific name of this class comes from the bristly tufts: Polychaeta means "many bristles."

Some polychaetes are sedentary and live in self-made tubes of sand and secretions; some burrow into sand and mud. Others, like fireworms (family Amphinomidae), crawl along the ocean floor searching for dead animal material, corals, crabs, snails, and other worms to eat. The hollow, sharp bristles on fireworm paddles help the animals grip the bottom and provide protection from predators.

Hawai'i hosts several species of fireworms. Some can extend and retract the bristles at the ends of their paddles. When the bristles are extended, the worms look like bristly caterpillars. Hawai'i's fireworms are from about 2 to 6 inches long and live under rocks and among coral rubble.

Several other kinds of polychaetes bear sharp jaws, for eating and for defending territory. In some parts of world, several types of jawed polychaetes grow large enough to inflict significant bites on humans. In Hawai'i, the largest worms grow to only an inch or so long; the smallest are barely visible to the naked eye.

Mechanism of Injury

Fireworm bristles are hollow, barbed, and sometimes filled with toxic fluid. The bristles break off easily in human skin on contact, introducing unidentified toxins into the wound.

Incidence

Fireworm stings are fairly common in Hawai'i. Some fireworms are brightly colored, inviting unwary divers to touch them. Others hide under rubble, sometimes stinging tide pool explorers who turn over rocks. One fireworm species lives on driftwood, eating gooseneck barnacles and stinging incautious beachcombers. Fireworms that eat dead marine life sometimes take baited hooks.

No polychaete worm bites have been reported in Hawai'i.

Prevention

Do not touch any bristly, segmented worm with your bare hands. If you are wearing gloves, be careful of touching fireworm-exposed areas of the gloves when taking them off. Wash the gloves after exposure.

Before handling rocks or driftwood, check for fireworms.

Anglers should use a stick, long tweezers, or other tool to remove a fireworm from a hook.

Signs and Symptoms

Fireworm stings most often cause a prickling or burning sensation lasting a few hours. Sometimes, the area either itches or feels numb. The pain usually decreases after several hours.

Fireworm bristles do not often pierce thickened, callused skin, but they easily penetrate soft skin. In those areas, a raised, red rash is common, sometimes with swelling and blisters. Rarely, the skin peels. Rashes may persist for two to three days. Discolored skin returns to normal in seven to ten days.

Careful examination of your skin after touching a fireworm may reveal silvery hairs sticking out of the skin. Sometimes, however, the offending fireworm bristles are not visible.

 First Aid

Remove visible bristles with tweezers. To remove tiny (often invisible) bristles, apply sticky tape to the area and pull it off gently. Regardless of technique, it is nearly impossible to remove all the bristles.

For persistent itching or skin rash, try 1 percent hydrocortisone ointment four times a day, and one or two 25-milligram diphenhydramine (Benadryl) tablets every six hours. These drugs are sold without prescription. Diphenhydramine may cause drowsiness: Do not drive, swim or surf after taking this medication.

No substance always relieves the pain and itching caused by these worms. Although unproven, home remedies of vinegar, rubbing alcohol, dilute ammonia, or a paste of meat tenderizer may help.

If the sting is severe, or if the area around any fireworm sting becomes infected, see a doctor. An allergic reaction to fireworm venom is always a possibility. Go to an emergency room at the first sign of hives, a feeling of overall illness, or breathing difficulty.

Advanced Medical Treatment

No specific antidote or clinically useful diagnostic tests exist for fireworm stings. After bristle removal, localized pain and rash are the primary problems. For severe inflammation not responsive to topical steroids and oral antihistamines, try oral steroids for several days.

For information about infected stings, see Part 2, *Staph, Strep,* and General Wound Care.

HYDROID STINGS

Hydroids (class Hydrozoa) grow on rocks, boat bottoms, and piers along tropical, subtropical, and temperate coastlines everywhere. These animals look like delicate seaweeds, but hydroids are actually animals related to corals, jellyfish, and sea anemones.

The hydroid *(Halocordyle disticha)* stings when it is handled. These hydroids are abundant in Kāne'ohe Bay and also can be found in the Ala Wai Small Boat Harbor, Kewalo Basin, Honolulu Harbor, and Ke'ehi Marina. Hydroids often grow on pilings, submerged lines, and boat bottoms. (William Cooke)

Hydroids are carnivores and use their nematocyst-laden feeding tentacles, positioned along "branches," to sting and catch passing shrimp, worms, and animal plankton.

Most hydroids have separate defense organs, often in the shape of tiny clubs, near their mouths. These clubs are well armed with both sticky and stinging nematocysts. The clubs defend the colony, and probably also help catch food.

Most hydroid colonies grow only a few inches tall. At least twenty-eight species inhabit the shallow waters of the main Hawaiian Islands.

Mechanism of Injury

Upon contact, hydroids' stinging nematocysts discharge venom into human skin. For details on nematocyst action, see Box Jellyfish Stings and Portuguese Man-of-War Stings.

Incidence

Hydroid stings are fairly common among people who clean fouled boat bottoms. Scuba divers sometimes get hydroid stings by accidentally brushing against a colony. In 1955, dislodged pieces of hydroid colonies growing on submerged rocks plagued construction men working on a pier in Hilo Harbor.[1]

Prevention

Wear protective clothing, especially on arms and hands, when cleaning boat bottoms.

When scuba diving, wear a body suit, gloves, and booties. If you are diving with bare skin exposed, be alert in surge areas where waves can push you against rocks and walls loaded with hydroids. Some hydroids form lovely, lacy colonies that invite inspection. Look, but do not touch.

Avoid walking on submerged rocks where you could break off pieces of hydroid colonies; these free-floaters sting on contact.

Signs and Symptoms

Most hydroid stings almost immediately produce small red bumps that remain itchy and painful for hours. Sometime victims feel a prickly sensation. The rash can last up to ten days.

Skin with hair on it (arms and legs) usually has less reaction than bare skin (wrists and ankles). More severe sting reactions are hives, blisters, and swelling.

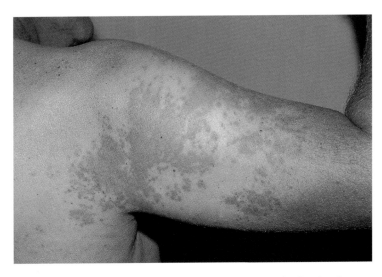

This diver broke out in a rash after brushing against a hydroid colony growing on a piling. The painful rash lasted several days. (Jerry Hughes, M.D.)

Stings to the eyes can cause pain, swelling, tearing, irritation, blurred vision, or light sensitivity.

✚ First Aid

Pick off any visible tentacles. Rinse the sting area thoroughly with water, salt or fresh, to remove any adhering nematocysts. Apply ice for pain relief.

For persistent itching or skin rash, try 1 percent hydrocortisone ointment four times a day, and one or two 25-milligram diphenhydramine (Benadryl) tablets every six hours. These drugs are sold without prescription. Diphenhydramine may cause drowsiness: Do not drive, swim, or surf after taking this medication.

In one study, alcohol, vinegar, and urine all discharged nematocysts from the hydroid *Lytocarpus philippinus,* common in the tropical Pacific but not found in Hawai'i. Seawater, fresh water, and distilled water did not discharge the nematocysts, nor did aloe vera or Stingose, a commercial spray sold for use on marine stings.[2] Heat, including sunburn, usually worsens hydroid irritation.[3]

Irrigate eye stings with copious amounts of room temperature tap water for at least fifteen minutes. If vision blurs or the eyes continue to tear, hurt, and swell or are light sensitive after irrigating, see a doctor.

If the skin rash worsens, pain persists, or allergic symptoms occur, or if a feeling of generalized illness develops, also see a doctor.

Advanced Medical Treatment

No specific antidote or clinically useful diagnostic tests exist for hydroid stings.

Unless a person is allergic, hydroid stings usually present no danger. Extensive stings unrelieved by antihistamines and topical steroids may be eased by oral steroids for several days. For information about infected stings, see Part 2, *Staph, Strep,* and General Wound Care.

No literature exists on hydroid stings to the eye. Eye stings from the sea nettle, a common Chesapeake Bay jellyfish, are usually self-limited, disappearing in twenty-four to forty-eight hours. One report recommends using topical steroids and cycloplegic drops. Some victims have persistent dilated pupils with blurred vision, particularly for nearby objects. Initially, intraocular pressure is elevated. This pressure remains high in some patients, producing glaucoma in the affected eye.[4]

LEATHERBACK *(LAI)* STINGS

Leatherback fish, or *lai (Scomberoides lysan),* belong to the same family as jacks *(ulua)* and scads *(akule).* Also known as queenfish, the silvery leatherback bears two rows of round spots along each side of the body. This species grows to about 28 inches long.

Near the dorsal (back) and anal (rear belly) fins of this fish lie several sharp, venomous spines. Some researchers believe other jacks also bear venomous spines. Only in leatherbacks, however, has venom apparatus been found.

Leatherback fish are members of the Jack family, generally called *ulua* in Hawai'i. Some researchers believe several jack species have venomous spines near their dorsal and anal fins. Only those of the leatherback, however, have been confirmed venomous. Island of Hawai'i. (Ricardo Mandojana, M.D.)

Leatherbacks cruise along Hawai'i's coastal waters, often above drop-offs, feeding on smaller fish and crustaceans. Juveniles venture far inshore, sometimes into brackish water, eating scales they tear off mullet and baitfish.

Hawai'i anglers regard leatherbacks as good gamefish, although the fish do not taste particularly good. They are known for their tough, leathery skin, hence their common name. Hawaiians used this durable skin for fashioning the heads on small coconut drums. Today, anglers make fishing lures of leatherback skin.

Mechanism of Injury

Leatherback fish have seven venomous dorsal spines and two venomous anal spines. A captured fish thrashing at the end of a line or on the deck of a boat can drive the spines into an angler's hand or foot, usually resulting in one or more puncture wounds.

Pain from these wounds is probably the result of tissue trauma, the effects of venom, and the introduction of slime and other surrounding substances into the wound.

Prevention

Anglers should wear gloves and boots while removing live fish from spears, lines, or nets. While cleaning a leatherback fish, keep your fingers away from the sharp spines near the dorsal and anal fins.

Use equal caution around other jackfish. They too may inflict painful wounds.

Incidence

No data are available on the frequency of leatherback puncture wounds in Hawai'i or elsewhere. Hawai'i anglers, however, have reported painful wounds (and bear scars) from the fins of several types of jacks.

Signs and Symptoms

Leatherback punctures can cause immediate pain out of proportion to the injury. Swelling and bleeding may accompany the injury.

 # First Aid

For pain relief, try soaking the area in hot, nonscalding water. This method has not been studied with wounds from leatherback fish, but is often effective with other fish stings, such as rays and scorpionfish. (Victims in pain may not be able to judge if the water is too hot; someone else should test the water temperature on his or her own hand to be sure it is not scalding.)

To treat minor wounds, gently pull the edges of the skin open and remove embedded material by rinsing or with tweezers. Then scrub directly inside the cut with gauze or a clean cloth soaked in clean, fresh water. Press on the wound to stop bleeding. If bleeding persists or if the edges of a wound are jagged or gaping, the victim probably needs stitches. Taping a cut shut is often effective but may leave a more visible scar than suturing. For more information about wound care, see Part 2, *Staph, Strep,* and General Wound Care.

Victims with any feeling of overall illness after a leatherback puncture should go directly to an emergency room.

Advanced Medical Treatment

No antivenom or diagnostic tests exist for leatherback fish envenomation.

For most patients, pain control, foreign body detection and removal, and localized wound care are the primary issues. Treat pain unrelieved by heat with buffered bupivacaine. Use soft-tissue x-ray or ultrasound to locate pieces of spine that may have broken off in the wound. Do not prescribe antibiotics for minor injuries with no sign of infection, except in immune-compromised patients. Discuss the risk of infection with the patient. For more information about wound care, see Part 2, Advanced Medical Treatment in *Staph, Strep,* and General Wound Care.

MANTIS SHRIMP CUTS

About three hundred species of mantis shrimp (order Stomatopoda) range throughout the world's tropical and subtropical oceans. Most live in burrows in the sand or mud, or in cracks and crevices of rocks and coral. These creatures grow from 1 to 12 inches long.

Mantis shrimp are named after the praying mantis, an insect famous for its large forelimbs. Like their namesake, mantis shrimp claws fold under their bodies like jackknives.

Many mantis shrimp feed on soft-bodied invertebrates and fish, lying motionless at the mouth of their burrows to ambush passing prey. When a fish or shrimp passes close by, the mantis shrimp strikes out

Hawai'i mantis shrimp *(Odontodactylus hawaiiensis)*. None of Hawai'i's mantis shrimps carry venom, but their lightning-fast striking claws can deliver a painful cut or bruise. (Colin J. Lau)

with its powerful forelegs, which are covered with barbed spines, spearing and capturing the prey.

Other species stalk clams, snails, and crabs. These mantis shrimp kill hard-bodied animals by smashing them with the heavy elbow of the unfolded foreleg. Other times, the shrimp drags its prey alive and whole to its burrow, where it breaks and eats it. Broken, empty shells lying on the ocean floor signal mantis shrimp burrows. Sometimes mantis shrimp disable crabs by breaking their claws. The predator then breaks open the crippled crab's shell to eat the meat.

Mantis shrimp use their powerful legs to defend themselves, as well as for feeding. The blows of some mantis shrimp are so powerful they have cracked aquarium glass.

Mechanism of Injury

Mantis shrimp either slash or punch their victims, depending on the species.

Incidence

These shrimp are common in Hawai'i. Because victims often do not know what hurt them, the rate of injury is unknown.

Prevention

Wear gloves while handling nets pulled from reefs. If a mantis shrimp is tangled in a net, remove it with a stick or tool, but never with your bare hand.

Underside of a mantis shrimp, *Odontodactylus* species, from the waters off Hale'iwa, O'ahu. Note the front claws folded under the body like jackknives. The shrimp lashes out with these claws to capture food and to defend itself. (Colin J. Lau)

If you are turning over stones or pieces of dead coral on the reef, wear gloves and turn the stone away from you, not toward you.

Be aware that mantis shrimp can lie buried in shoreline mud. A surprise cut or bruise to your foot or hand may well be from a striking mantis shrimp.

Signs and Symptoms

A strike from one of these shrimp causes immediate pain. Often, a mantis shrimp slashes the skin open, but some species punch, resulting in a bruise.

 # First Aid

For minor wounds, scrub directly inside the cut with gauze or a clean cloth soaked in clean, fresh water. Press on the wound to stop bleeding. If bleeding persists or if the edges of a wound are jagged or gaping, the victim probably needs stitches. Taping a cut shut is often effective but may leave a more visible scar than suturing. For more information about wound care, see Part 2, *Staph, Strep,* and General Wound Care.

Banded mantis shrimp
(Lysiosquilla maculata).
Anglers often catch these
mantis shrimp near
Kāne'ohe (He'eia) Fishing
Pier, Kāne'ohe Bay. (Hawaii
Fishing News photo)

 Advanced Medical Treatment

Hawai'i's mantis shrimp do not carry venom. Injuries are generally minor.

Thoroughly scrub, explore, irrigate, and debride all lacerations. Do
not prescribe antibiotics for minor injuries with no sign of infection,
except in immune-compromised patients. Discuss the risk of infection
with the patient.

Suturing wounds to control bleeding, preserve function, or improve
appearance is appropriate.

For more information, see Part 2, Advanced Medical Treatment in
Staph, Strep, and General Wound Care.

MORAY EEL *(PUHI)* BITES

Moray eels, or *puhi* (family Muraenidae), are fish with long, narrow
bodies ideal for taking shelter in holes in the reef. At least thirty-five
species of moray eels are found in Hawaiian waters, making them one
of Hawai'i's most abundant, if not visible, fish.

Though not often seen, the stout moray *(Gymnothorax eurostus)* is one of the most common inshore moray eels in Hawai'i. It grows to at least 22 inches. Hanauma Bay. (D. R. and T. L. Schrichte)

The majority of Hawai'i's morays are small, only about 2 feet long when fully grown. Large eels, growing to about 6 feet, represent a small minority. One species, the giant moray *(Gymnothorax javanicus)*, grows to a confirmed 8 and possibly 10 feet long. This largest of all morays is rare in Hawaiian waters.

Like many fish, moray eels breathe by pumping water in through their mouths and out through the gills. The resulting opening and closing of the mouth bares the teeth, giving morays a menacing appearance.

Some moray eels have backward-tilting, needlelike teeth, ideal for grabbing fish, octopus, and crabs. The yellowmargin moray, or *puhi paka (Gymnothorax flavimarginatus),* has a particularly bad reputation for being fierce. This may be because the *puhi paka* is one of the most common of Hawai'i's large, sharp-toothed eels, and therefore is encountered frequently.

Several kinds of moray eels, such as the whitemargin moray *(Gymnothorax albimarginatus),* have serrated teeth. Bites from these morays reportedly hurt more than the injury warrants, suggesting the presence of venom. The fish, however, has no venom apparatus. Another serrated-toothed moray, the yellowmouth moray *(Gymnothorax nudivomer),* has toxic mucus covering its body.[1]

Several morays, such as the zebra moray *(Gymnomuraena zebra),* have pebble-shaped teeth, efficient for crushing crabs and other shelled animals.

Mechanism of Injury

Sharp-toothed moray eels can inflict deep puncture wounds. Their long teeth can introduce bacteria deep into the wounds, often causing infections.

Bites occur most often on hands or feet, where nerves and tendons lie close to the surface. If a bite victim pulls away abruptly, moray teeth create slash wounds that sometimes cut nerves, tendons, ligaments, and blood vessels. Eel teeth often break off and lodge inside wounds.

Although some researchers suspect that serrated-toothed moray eels deliver venom with their bites, no one has identified venom in eel mouths. The significance of the yellow-mouth moray's toxic skin mucus in human bites is unknown.

Pebble-toothed morays can crush fingers or toes.

Eating moray eels can cause ciguatera fish poisoning. For more information, see Part 2, Ciguatera Fish Poisoning.

Incidence

Moray eels have undeserved reputations for viciously attacking people. These fish usually retreat into holes and hide when someone approaches them. Bites sometimes result when a diver hunting lobsters or shells accidentally places a hand near a moray's mouth, and even then, the eel does not always attack. Because people and moray eels are abundant in Hawai'i, bites are relatively common.

This whitemouth moray eel in Hanauma Bay (Gymnothorax meleagris) opens wide, allowing a Hawaiian cleaner wrasse (Labroides phthirophagus) to browse inside its mouth. Whitemouth morays grow to about 3 1/2 feet. Whitemouth and stout morays alike have three inner rows of fanglike teeth in addition to the outer row, plus teeth at the side of each jaw. (D. R. and T. L. Schrichte)

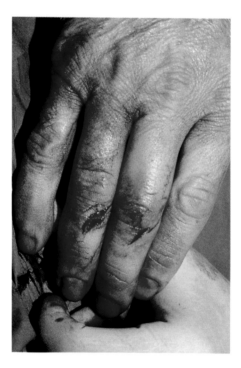

These finger wounds resulted from a moray eel bite. Slash wounds like these are typical in bites of species bearing sharp teeth. (Ricardo Mandojana, M.D.)

Prevention

Do not feed moray eels or dive near people who do. Some eels that have become accustomed to being fed have bitten and seriously injured unsuspecting divers.

Snorkelers and divers should check for eels before grabbing onto rocks for hand-holds. Aquarists with captive morays should wear a heavy glove when immersing a hand in the tank. Surfers and wind-surfers off their boards should not stand on the reef. Wearing reef shoes helps protect feet.

Anglers must use caution when spearing morays or trying to remove them from lines or nets. While struggling for its life, a moray eel may become extremely aggressive.

If a moray eel does bite, jerking your hand or foot backwards may cause the fish's sharp teeth to make deep, slashing cuts. If you are bitten, try to let the eel open its mouth by itself, which it almost always will do. Stories about decapitating moray eels or prying their jaws open with a knife to release a victim have not been verified in Hawai'i.

Signs and Symptoms

A moray eel bite usually looks like two puncture wounds or two equally long, deep cuts.

Not all moray eels have sharp teeth. The zebra moray *(Gymnomuraena zebra)* has round, pebblelike teeth, which it uses to grind up invertebrates. The teeth can bruise a person's fingers or toes. Shark's Cove. (D. R. and T. L. Schrichte)

In 1873, a Lāna'i resident wrote that the *puhi paka* (yellowmargin moray) would "take off a toe or snap off an exposed naked foot if he gets the chance."[2] No one has recorded moray eel bite injuries like these in Hawai'i. That the amputations ever happened is questionable, since moray eel teeth are adapted for grasping moving prey rather than cutting it into pieces.

 # First Aid

Moray eel bites are notorious for tendon and nerve damage in hands and feet.

For minor bites, gently pull the edges of the skin open and remove any embedded teeth by rinsing or with tweezers. Then scrub directly inside the bite with gauze or a clean cloth soaked in clean, fresh water. Press on the wound to stop bleeding. If bleeding persists or if the edges of a wound are jagged or gaping, the victim probably needs stitches. Taping a bite shut is often effective but may leave a more visible scar than suturing. For more information about wound care, see Part 2, *Staph, Strep,* and General Wound Care.

If a finger or toe is numb or will not move normally after a bite, see a doctor immediately.

Victims who appear pale, sweaty, and nauseated are in danger of fainting. Lower the victim to the ground.

 # Advanced Medical Treatment

Some moray eel bites may contain venom. No specific treatment exists for these bites.

Moray teeth often break off inside wounds and may be visible on soft-tissue x-ray or ultrasound. Remove them to avoid infection or foreign body granulomas. Thoroughly scrub, explore, irrigate, and debride all wounds. Examine for tendon and nerve damage and repair, or refer, as necessary. Suturing wounds to control bleeding, preserve function, or improve appearance is appropriate.

Do not prescribe antibiotics for minor injuries with no sign of infection, except in immune-compromised patients. For more information about antibiotic therapy, see Part 2, *Staph, Strep,* and General Wound Care.

NEEDLEFISH *('AHA)* PUNCTURES

Needlefish, or *'aha* (family Belonidae), have narrow, tubular bodies ending in rigid, pointed jaws that make up about a quarter of the fish's length. Both jaws bear sharp teeth, for catching fish. The needlefish strikes at a fish with a sideways movement of the head, then swallows it whole.

Like their distant relatives the flying fish, or *mālolo* (family Exocoetidae), needlefish can leap from the water at up to 38 miles an hour, skimming the surface for a few yards before falling back to the water. At night, lights tend to attract and excite the fish, often causing them to jump at speed. Needlefish beaks have penetrated the wooden hulls of outrigger canoes.

Needlefish are common throughout all tropical waters. Four species, all similar in appearance, inhabit Hawai'i's reefs. Most needlefish swim

Needlefish, or *'aha,* are capable of great leaps that can accidentally drive their sharp beaks into human flesh. Although injuries from needlefish are rare, some can be life threatening. Pictured here is a crocodile needlefish *(Tylosurus crocodilus)* from Waimea Bay. (Ricardo Mandojana, M.D.)

in small schools, but some are solitary. Hawai'i's largest species grows to about 3 feet long.

It is easy to miss seeing nearby needlefish while snorkeling or diving, because they hover near the surface, blending well against their backgrounds. The fish are silvery below, bluish above.

Mechanism of Injury

When needlefish leap from the water, their remarkably sharp jaws can drive into human tissue like an ice pick. The fish beaks are narrow enough to slide between ribs and vertebrae, and long and rigid enough to puncture organs, blood vessels, and even the brain. All or part of the beak may break off and remain in the wound.

Incidence

One death has been documented in Hawai'i from a needlefish puncture. In 1977, a leaping needlefish struck a ten-year-old Kaua'i boy near the right eye while he was night fishing in a boat with his father. The beak penetrated the boy's brain, killing him.

In other parts of the Pacific, needlefish beaks have punctured people in the head, neck, arms, legs, chest, and abdomen. One death occurred from a punctured thoracic aorta. Another victim bled to death from a neck puncture. Needlefish have also caused spinal cord injuries and brain damage. For many village fishermen in the Pacific, needlefish are a greater occupational hazard than sharks.

At greatest risk are night reef fishermen carrying lights in low boats. Japanese lobster fishermen holding dive lights have been punctured. Several swimmers have been seriously injured: one in New Zealand, one in Japan, another in the Red Sea. In separate incidents, two Solomon Islanders were blinded in one eye from needlefish punctures.

Prevention

At night, scuba divers should leave their lights off until well submerged.

Night fishermen either holding lights or carrying them in small boats should be aware that needlefish punctures can cause life-threatening injury. Do not ignore these punctures, even if they seem minor.

Signs and Symptoms

The surface wound of a needlefish puncture can look deceptively small and trivial. It is possible, however, that deep penetration of the fish's long beak has done serious damage.

Be alert for symptoms of damage to deep tissue or organs, such as severe pain, difficulty breathing, or excessive bleeding.

First Aid

Do not try to remove a needlefish beak from the flesh. If the fish is still attached to its embedded beak, cut it off. Take the victim to an emergency room for removal of the beak and evaluation of possible deep tissue damage.

If a needlefish pulls its beak from the flesh or if the beak breaks off under the skin, and the wound bleeds, press gently but firmly over the wound to control bleeding.

Needlefish wounds are at high risk for retained beak, serious internal injury, and infection. Seek medical consultation for these wounds.

Advanced Medical Treatment

Needlefish bear no toxins. Punctures can be deep and life threatening even though the entry site is small. (One needlefish beak was removed from a thoracic aorta at autopsy.)[1]

Regard needlefish punctures as stab wounds. When the patient is stable, use soft-tissue x-ray or ultrasound to locate a possible retained beak. In Israel recently, standard cervical spine films missed a 1 1/2-inch beak fragment in a woman's neck. The beak was found one month later with soft-tissue x-ray and computed tomography.[2]

Removal of the beak, and cleaning and debridement of the wound, often requires surgical intervention. For trauma victims in hemorrhagic shock, use the airway, breathing, and circulation (ABCs) protocol of trauma management.

For information about care of uncomplicated wounds, see Part 2, *Staph, Strep,* and General Wound Care.

OCTOPUS *(HE'E)* BITES

An octopus catches prey by pouncing on it and enclosing the prey in the web that stretches between its eight arms. The octopus immobilizes its catch by biting with two parrotlike jaws, its beak. The bite delivers a paralyzing venom from the animal's salivary glands; the octopus carries the prey to its den, where it dismembers and eats it.

Wearing gloves and being gentle helps this diver handle a night octopus, or *he'e pūloa (Octopus ornatus),* without getting bitten. Most octopus bites are reported from fishermen, who hunt the octopuses for food. Hanauma Bay. (D. R. and T. L. Schrichte)

In Hawai'i, octopuses are called *he'e* (Hawaiian), *tako* (Japanese), or squid (local vernacular). Three species inhabit Hawaiian reefs. One is the day octopus, or *he'e mauli (Octopus cyanea).* This octopus is dusky gray or tan and hunts for crabs and shrimp during the day on exposed areas of the reef. The day octopus grows to about 2 feet long from the top of its head to the end of its outstretched arms.

The reddish brown night octopus, or *he'e pūloa (Octopus ornatus),* hunts on the reef at night. This nocturnal octopus, easy to identify by its white spots, is a bit smaller and thinner than the day octopus.

Hawai'i's third species is the crescent octopus, named by the student who recognized it twenty years ago. No scientific name has yet been assigned to the creature, which looks like a small day octopus.

Two species of blue-ringed octopuses ranging throughout Australia, New Zealand, New Guinea, and Japan produce a venom containing the potentially lethal tetrodotoxin. Blue-ringed octopuses *(Hapalochlaena maculosa* and *H. lunulata)* are the only species known to inflict fatal bites on humans. These octopuses do not inhabit Hawaiian waters.

Mechanism of Injury

An octopus bite can tear the skin, sometimes producing bleeding. The octopus sometimes injects venom from its salivary glands when biting humans. Octopus venom contains enzymes that break down proteins, and a glycoprotein (sugar plus protein) toxin.

Hawai'i's octopuses all carry venom, but none contain tetrodotoxin.

Incidence

Local fishermen report that most bites come from night octopuses. Typically, fishermen wade onto shallow reef flats, either spearing or catching octopuses by hand. In their death struggle, octopuses sometimes try to bite the hand that holds them. To subdue these writhing, mucus-covered creatures, some fishermen bite the octopus between the eyes.

Hawai'i divers usually handle octopuses without being bitten. If the animal is handled gently, it rarely bites. Octopus bites are usually the result of someone harassing the animal.

Prevention

To avoid octopus bites, do not take the animals out of the water. In the water, do not antagonize them. If you handle an octopus in the water, wear gloves and be kind. Better yet, just watch and do not touch.

Signs and Symptoms

An octopus bite usually looks like two puncture wounds. If the animal injects venom, the pain is similar to a bee sting, with a tingling or pulsating sensation around the wound. Pain may radiate to include the entire arm or leg. Venomous octopus wounds can bleed profusely. Redness and swelling of the affected area is common. Some victims experience intense itching around the wound.

Octopuses have two powerful, beaklike jaws (the black objects) that can tear and bite off pieces of tissue. A tonguelike radula (the brownish, rasplike object between the black jaws) then pulls the tissue into the octopus' mouth. United Fishing Agency Auction, Honolulu. (D. R. and T. L. Schrichte)

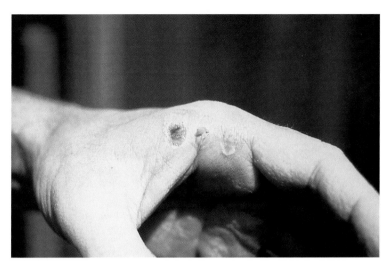

A classic wound from an octopus bite. (World Life Research Institute)

 # First Aid

Unless a person is allergic to it, venom produced by Hawai'i's octopuses is not life threatening.

To clean an octopus bite, gently pull the edges of the wound open. Then scrub directly inside the bite with gauze or a clean cloth soaked in clean, fresh water. Press on the wound to stop bleeding. If bleeding persists or if the edges of a wound are jagged or gaping, the victim probably needs stitches. Taping a bite shut is often effective but may leave a more visible scar than suturing. For more information about wound care, see Part 2, *Staph, Strep,* and General Wound Care.

 # Advanced Medical Treatment

No antidote or specific tests exist for octopus envenomation. In Hawai'i, octopus bites are not usually dangerous. Octopus venom does not require specific therapy.

Thoroughly scrub, explore, irrigate, and debride all wounds. Examine for tendon and nerve damage and repair, or refer, as necessary. Suturing wounds to control bleeding, preserve function, or improve appearance is appropriate.

Do not prescribe antibiotics for minor injuries with no sign of infection, except in immune-compromised patients.

For more information about antibiotic therapy, see Part 2, *Staph, Strep,* and General Wound Care.

PORTUGUESE MAN-OF-WAR (PA'IMALAU) STINGS

The Portuguese man-of-war, or *pa'imalau* (*Physalia* species), is related to jellyfish. This creature floats on the surface of tropical and subtropical waters throughout the world by a bluish purple, gas-filled sac. On top of the sac lies a crest. During periods of calm weather, the crest flattens and the animal drifts with offshore currents. When the winds blow, the Portuguese man-of-war raises its crest and sails at a 45-degree angle to the wind.

Tentacles bearing batteries of stinging cells called nematocysts hang beneath the animal's float. When the tentacles come in contact with a fish or drifting animal, the nematocysts discharge, entangling and stinging the prey. The tentacles pull the immobilized prey to the man-of-war's mouth parts (also hanging under the float), which expand to swallow the meal. Some researchers divide Portuguese man-of-war into two species, *P. physalis* and *P. utriculus*. The two differ in appearance but may be two forms of the same species.

The creature described as *P. physalis* has a float up to 12 inches long and multiple tentacles reaching a length of nearly 100 feet. Each tentacle bears up to 750,000 nematocysts. This sometimes lethal creature is common in the Western Atlantic and was reported in Australia in 1991. *P. utriculus* is smaller, with floats up to about 2 inches long and a

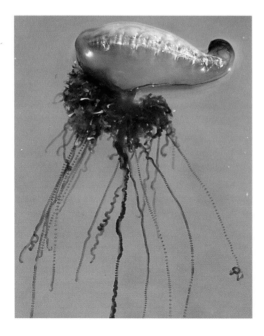

Portuguese man-of-war are open-ocean relatives of jellyfish. These wind-driven creatures show up in Hawai'i's bays and on the beaches during strong wind conditions. Bellows Beach. (D. R. and T. L. Schrichte)

single tentacle up to about 30 feet long (sometimes with several thinner, less obvious tentacles). Although this is the only species officially reported in Hawaiian waters, larger Portuguese man-of-war with multiple tentacles have washed up on Hawai'i's beaches.

Mechanism of Injury

Portuguese man-of-war sting unsuspecting swimmers with the same stinging cells, or nematocysts, the creatures use to catch food. This mechanism of action is similar for all nematocysts. (See Box Jellyfish Stings.) Even though different nematocysts have similar stinging mechanisms, they are by no means alike. Nematocysts are distinct to each species. Some animals even have different types of nematocysts on the same tentacle. This diversity could explain why certain treatments, such as vinegar dousing, works for one type of sting (box jellyfish) and not another (Portuguese man-of-war).

Unlike box jellyfish, Portuguese man-of-war are visible on the water's surface. Their long, stinging tentacles trail beneath the wave tops, making swimmers and surfers particularly vulnerable.

Nematocyst venom contains an impressive number of active compounds, both protein and nonprotein. Chemical transmitters similar to those that trigger nerve conduction, cardiac contraction, and allergic reactions are among these compounds.

No one is immune to nematocyst venom. Some people are allergic. Delayed skin allergies are common.

Incidence

Honolulu city and county lifeguards estimate nearly 100 percent of the stings they treat at windward and North Shore beaches come from Portuguese man-of-war. Estimates of Hanauma Bay stings are about 50 percent Portuguese man-of-war and 50 percent box jellyfish. Except in Kona weather, box jellyfish cause most leeward coast stings.

In 1994, lifeguards treated approximately 6,500 Portuguese man-of-war sting victims. Of these, eight were major stings, requiring additional medical treatment.

Portuguese man-of-war are wind-driven creatures, commonly appearing in Hawai'i's bays and beaches during periods of strong onshore winds. The winds blow onto east and north shores during trade wind conditions (the majority of the time), and onto south and west shores during the infrequent periods of Kona, or south winds.

No deaths have been reported in Hawai'i from Portuguese man-of-war stings. Some victims, though, suffer serious overall illness or have severe allergic reactions from stings.

Portuguese Man-of-War Stings on O'ahu, 1994

Beach	Minor Stings	Major Stings
Hanauma Bay	86	12
Sandy	102	0
Makapu'u	10	0
Waimānalo (includes patrols on all-terrain vehicles)	3,888	2
Bellows (weekends and holidays only)	1,340	3
Kailua	960	3
Kualoa	249	0
Keawa'ula	14	1
Ali'i	8	0
Sunset	2	0
'Ehukai	3	0
Ke Waena	1	0
Waimea	3	0

Three people have died in the United States from Portuguese man-of-war stings, all from the large species, *P. physalis*. All three victims were stung in coastal waters off the Southeast.

Prevention

Portuguese man-of-war are surface floaters. Swimmers and surfers should watch for the bluish floats, giving them wide berth to avoid the creatures' trailing tentacles. Wearing wet suits or synthetic body suits in the ocean during windy weather helps cut down on sting incidence. Heed beach postings and lifeguard warnings.

Scuba divers should look up while ascending, surfacing with one arm outstretched to protect their faces.

Signs and Symptoms

Symptoms of Portuguese man-of-war stings vary from person to person. Reactions depend on how much skin was exposed and for how long, and the size and sensitivity of the victim. Because children and small adults receive more venom per pound of body weight than larger

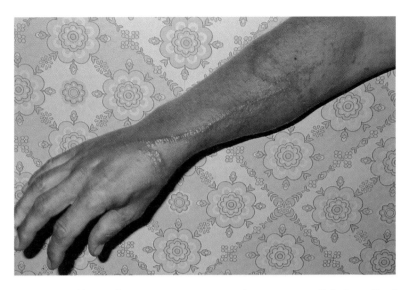

A two-day-old sting from a Portuguese man-of-war on Australia's Great Barrier Reef. A dive operator immediately poured vinegar on the sting, with no obvious effect. The victim reported that for about two hours the pain felt like "glowing, red-hot wires on my arm." (Walter Koestenbauer)

people, they tend to have more intense reactions. Also, the body part injured can determine a person's reaction. The eyes and face are usually more sensitive than the arms and legs. Stings to the eyes can cause pain, swelling, tearing, irritation, blurred vision, or light sensitivity.

In many victims, Portuguese man-of-war stings leave no visible marks on the skin. Other people exhibit symptoms that range from slight redness to hivelike welts and blisters. Rarely, a rash spreads to the entire body.

The classic Portuguese man-of-war sting in Hawai'i causes an immediate burning sensation, with slight reddening where the tentacles touched. The pain and redness disappear after twenty minutes to an hour with no treatment. The sting of a Portuguese man-of-war is usually less painful and of shorter duration than that of a box jellyfish.

Lymph nodes in the vicinity of the sting sometimes swell. This is usually a sign of the toxic effect of the venom, or it may be an early sign of infection.

Rarely, a sting will cause nerve damage, resulting in numbness or weakness of the affected area for weeks to months. Occasionally, a scar or permanent discoloration of the skin appears at the site.

Severe allergic reactions can be life threatening. Signs of the reaction are confusion, irregular heartbeat, difficulty breathing, high or low blood pressure, and weakness.

Delayed allergic-type skin reactions can appear weeks later and persist for months. Usually they itch but do not hurt.

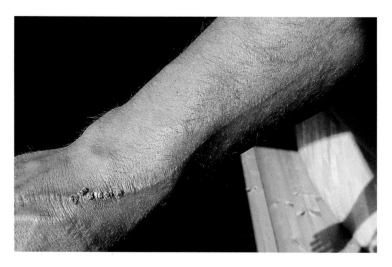

The same sting sixteen days later, in Graz, Austria, the victim's home. At twenty-one days, the sting lines were still raised and visible. (Walter Koestenbauer)

✚ First Aid

Follow these steps for treating the average Hawai'i Portuguese man-of-war sting:

1. Pick off any visible tentacles with a gloved hand, stick, or anything handy, being careful to avoid further injury.
2. Rinse the sting thoroughly with salt or fresh water to remove any adhering nematocysts.[1]
3. Apply ice for pain control.[2]
4. Irrigate eye stings with copious amounts of room temperature tap water for at least fifteen minutes. If vision blurs or the eyes continue to tear, hurt, and swell or are light sensitive after irrigating, see a doctor.
5. For persistent itching or skin rash, try 1 percent hydrocortisone ointment four times a day, and one or two 25-milligram diphenhydramine (Benadryl) tablets every six hours. These drugs are sold without prescription. Diphenhydramine may cause drowsiness: Do not drive, swim, or surf after taking this medication.

Although formerly thought to be effective, vinegar is no longer recommended for Portuguese man-of-war stings. In a laboratory experiment, vinegar dousing caused discharge of nematocysts from the larger (*P. physalis*) man-of-war species. The effect of vinegar on the nematocysts of the smaller species (which has less severe stings) is mixed: vinegar inhibited some and discharged others.[3]

No studies support applying heat to Portuguese man-of-war stings.

Studies on the effectiveness of meat tenderizer, baking soda, papain, or commercial sprays (containing aluminum sulfate and detergents) on nematocyst stings have been contradictory. These substances may cause further damage. In one United States fatality from a Portuguese man-of-war, lifeguards had immediately sprayed papain solution on the victim's sting. Within minutes the woman was comatose, and later she died.[4]

Alcohol and human urine may be harmful on Portuguese man-of-war stings. An Australian study reports that alcohol and urine caused massive nematocyst discharge in the box jellyfish *Chironex fleckeri,* also known as the sea wasp.[5]

Most Hawai'i Portuguese man-of-war stings disappear by themselves, sometimes within fifteen or twenty minutes. Because of this, even harmful therapies often appear to work. A key concept in the first aid of any injury is: Do no harm. Therefore, avoid applying unproven, possibly harmful substances on stings.

See a doctor if pain persists, the rash worsens, a feeling of overall illness develops, a red streak develops between swollen lymph nodes and the sting site, or if either area becomes red, warm, and tender. For more information about signs of infection, see Part 2, *Staph, Strep,* and General Wound Care.

Few Portuguese man-of-war stings in Hawai'i cause life-threatening reactions, but this is always a possibility. Some people are extremely sensitive to the venom; a few have allergic reactions. Regard even the slightest breathing difficulty or altered level of consciousness as a medical emergency. Call for help and use an automatic epinephrine injector if one is available. (See Part 2, Allergic Reactions.)

Advanced Medical Treatment

No specific antidote or clinically useful diagnostic tests exist for Portuguese man-of-war stings. Application of vinegar and heat may make these stings worse. Serious reactions occur, but are rare. No fatalities have been reported in Hawai'i.

Portuguese man-of-war patients may need oral or IV narcotics for pain control.

Rarely, the skin blisters. Treat these patients for partial-thickness burns.

Recurrent dermatitis with a painless, itching rash sometimes accompanies Portuguese man-of-war stings. Therapy is unproven. For persistent cases, try oral steroids for several days.

A few people have anaphylactic reactions to Portuguese man-of-war stings. These patients require standard therapy of epinephrine, antihistamines, rehydration, and airway support. (See Part 2, Allergic Reactions.)

A severe reaction to some nematocyst venom is cardiac toxicity presenting as arrhythmias, hypotension, and conduction delay. The symptoms may be caused by the toxin's effect on ion channels. Most of these victims respond to IV fluids. Calcium channel blockers during resuscitation, however, may be beneficial. Calcium may increase cardiac toxicity.[6]

No reports exist on Portuguese man-of-war stings to the eyes. In eye stings from sea nettles (jellyfish common in Chesapeake Bay), the injury is usually self-limited, disappearing in twenty-four to forty-eight hours. One study recommends using topical steroids and cycloplegic drops. Some victims have persistent dilated pupils with blurred vision, particularly for nearby objects. Initially, intraocular pressure is elevated. This pressure remains high in some patients, producing glaucoma in the affected eye.[7]

Several cases of motor and sensory peripheral nerve deficits have been reported after jellyfish stings. All developed several days after the sting, affected nerves near the sting site, were not associated with impaired circulation, and improved over several months without treatment.[8]

RAY STINGS

Rays are relatives of sharks. Rays and sharks have skeletons of cartilage instead of bone and gill slits instead of bone-covered gill openings; and they lack the air sacs, called swim bladders, that most fish

A sting ray raises its spine as a bather approaches. (Ricardo Mandojana, M.D.)

use for buoyancy. Here the resemblance ends. Rays have flat bodies with extended, winglike fins. They are gentle, shy fish, with neither shark teeth nor shark appetites.

Three types of rays inhabit Hawai'i's waters: manta rays, or *hāhālua* (family Mobulidae); eagle rays, or *hailepo* (family Myliobatidae); and stingrays, or *hīhīmanu* (family Dasyatidae). Manta rays are by far the largest of all rays, growing up to 20 feet across and weighing up to 3,000 pounds. These plankton eaters swim forward with their mouths open, sifting the water for food. Manta rays have no stinging spines on their tails.

Hawai'i's only eagle ray, the spotted eagle ray *(Aetobatus narinari),* grows to about 6 feet across. It cruises the ocean floor looking for clams, snails, and other mollusks. Eagle rays occasionally swim in groups, in graceful formation. The spotted eagle ray has several venomous spines at the base of its tail, used for defense only.

Hawai'i hosts three kinds of stingrays, all growing 3 to 4 feet wide. Two, the Hawaiian stingray *(Dasyatis brevis)* and the slightly smaller brown stingray *(Dasyatis latus),* cruise along the ocean floor looking for clams and snails. When not actively hunting, these stingrays, dark on top and light below, often bury themselves in sand, hiding while resting.

The third stingray in Hawai'i is the pelagic stingray *(Dasyatis violacea),* found offshore in open water rather than on the ocean floor. The pelagic stingray has a dark back, but unlike the bottom-dwelling rays, its underside is dark grayish purple. This stingray eats fish, squid, and crustaceans.

All three stingrays have one or more venomous spines on the back of their tails. Like their eagle ray relatives, stingrays use the spines only in defense.

Mechanism of Injury

Ray stingers are flat, pointed, bonelike spines with backward-pointing barbs lining the edges. Along the underside of each spine lie two grooves containing glandular, venom-producing tissue. The spine and its glandular tissue is covered by a layer of skin.

When a ray is startled or caught, the fish whips its tail around, driving its serrated-edged spine into the intruder. During this process, the skin covering the spine often breaks, releasing venom and venom gland tissue into the wound. Envenomation takes place in about two-thirds of reported ray injuries. Often, pieces of the ray's skin and spine also remain in the wound.

Occasionally, the entire spine breaks off the ray's tail and lodges in the wound. Because backward-pointing barbs line the spines, removing them often tears tissue. Ray wounds can be straight cuts or puncture wounds.

Accidentally stepping on a sting ray is the typical way swimmers get stabbed with the creature's venomous barbs. (Ricardo Mandojana, M.D.)

Stingray venom consists of a mixture of proteins, some heat-sensitive. The venom may directly affect heart muscle.

Incidence

Manta rays pose no threat to humans. Eagle rays pose little threat to waders, since their stingers are located at the base of the tail and therefore are not well adapted as striking organs. Stingray spines are located farther out on the tail, making them efficient weapons. Hawai'i's stingrays are not common on the reef, so wounds from these fish are rare.

Estimates of stingray injuries in U.S. coastal waters are approximately two thousand a year. Most of these injuries involve people who wade in shallow water. Anglers are at risk whenever they are landing any living eagle ray or stingray.

Most wounds occur on people's feet and legs. Puncture wounds to the chest and abdomen have caused fatalities in other parts of the world. At least two deaths in Australia and one in New Zealand were from stingray spines penetrating the chest. Other deaths have been associated with massive infection and tetanus as a result of poor wound care. No deaths from rays have been reported in Hawai'i.

Prevention

When you are walking in water over a sandy bottom, shuffle your feet to startle rays into swimming away. Stingray spines can easily penetrate rubber or neoprene shoes. Stingrays have punctured the sides of wooden boats.

When you are swimming or diving, never chase or try to ride any ray.

Fishermen should remove rays from nets by using a tool such as a boathook, especially in an enclosed space. Instead of landing a hooked ray onto the deck of a boat, cut the line and let it go. Even a half-dead ray has the potential of whipping its tail in defense, causing a serious or even fatal injury.

Signs and Symptoms

If a ray spine enters a victim but does not penetrate sufficiently to break the spine's skin, the wound looks and feels like any other puncture wound or cut.

Ray wounds in which envenomation takes place cause immediate, intense pain, swelling, and usually bleeding. Pain peaks after thirty to ninety minutes and is gone in six to forty-eight hours. At first, the victim feels pain within about 4 inches of the wound, but the pain gradually expands. The wound area first looks bluish, then turns red.

Symptoms of overall illness from ray venom are weakness, nausea, vomiting, diarrhea, sweating, dizziness, fast heartbeat, headache, seizures, muscle pain and cramps, paralysis, low blood pressure, and irregular heartbeat.

 # First Aid

For ray wounds without envenomation, clean by gently pulling the edges of the skin open and scrub directly inside the cut with gauze or a clean cloth soaked in clean, fresh water.

For ray wounds with venom release (evident by severe pain), rinse the area immediately with whatever water is handy (ocean water if fresh water is not available) to remove poison gland tissue and venom. Remove any parts of an embedded spine with tweezers and thorough scrubbing and rinsing.

Press on the wound to stop bleeding. If bleeding persists or if the edges of a wound are jagged or gaping, the victim probably needs surgery to trim, clean, and repair the wound. For more information about wound care, see Part 2, *Staph, Strep,* and General Wound Care.

For pain control after envenomation, soak the area in hot, nonscalding water for thirty to ninety minutes.[1] (Victims in pain may not be able to judge if the water is too hot; someone else should test the water temperature on his or her own hand to be sure it is not scalding.) Often, when the water cools, the pain returns. Repeat hot water soaks for up to two hours. After two hours, heat is of little value.

Ray wounds often become infected, and some people have reactions to the venom. See a doctor for redness, swelling, or delayed healing. Victims with a finger or toe that is numb or will not move normally, or anyone with a feeling of general illness after a ray sting, should go directly to an emergency room.

Advanced Medical Treatment

No antivenin or diagnostic tests exist for ray envenomations. In most cases, controlling pain, finding and removing the spine and its venomous tissue, and treating the wound are the primary issues.

Some toxins in ray venom are heat labile. Inactivate them with hot water soaks as detailed above in the First Aid section. If pain persists, treat with buffered bupivacaine. Neither antihistamines nor steroids are beneficial in treating pain and inflammation from ray stings.

Ray spines, with their glandular venomous tissue, often break off in the wound. Spine fragments not visible in regular x-ray may be seen in soft-tissue x-ray or by ultrasound. Backward barbs and their fragments can be extremely difficult to detect and remove. Surgical consultation is sometimes necessary for spine removal, wound debridement, or for punctures to the neck, chest, or abdomen. Some practitioners recommend excision and packing of ray wounds.[2]

Using antibiotics to decrease the risk of wound infection is unproven. Reserve antibiotics for victims with damaged immune systems or severe puncture wounds. For more information about antibiotic therapy, see Part 2, *Staph, Strep,* and General Wound Care.

Rarely, this venom has systemic effects, including weakness, vomiting, and cardiac arrhythmias. Use standard antiarrhythmic therapies. Observe envenomated patients for at least four hours to watch for systemic reactions.

SCORPIONFISH *(NOHU)* STINGS

Scorpionfish, or *nohu* (family Scorpaenidae), get their name from the venomous spines they carry on their dorsal (back), pelvic (belly), and anal (rear belly) fins. The most notorious stinging fish, the stonefish (family Synanceiidae), are not found in Hawai'i.

Hawai'i hosts twenty-five species of scorpionfish, including the lionfish, or *nohu pinao (Dendrochirus barberi),* and turkeyfish, also *nohu pinao (Pterois sphex).* These graceful, colorful fish are popular in the aquarium trade.

Hawai'i's scorpionfish, such as this *nohu 'omakaha* at the Waikīkī Aquarium, have venom-tipped spines on their backs and bellies. Punctures from these fish are painful but are not life threatening. (Susan Scott)

Other scorpionfish are regarded as good food fish in Hawai'i: the devil scorpionfish, or *nohu 'omakaha (Scorpaenopsis diabolus),* and the titan scorpionfish, or *nohu (Scorpaenopsis cacopsis).*

The titan scorpionfish is Hawai'i's largest, growing to about 20 inches long. These scorpionfish (and most others) blend well with their surroundings and are almost impossible to see, even when resting right in front of a diver's face.

All scorpionfish are carnivores. The bottom-dwelling, camouflaged species lie motionless, waiting for an unsuspecting fish or shelled animal to pass nearby. With a quick lunge, the scorpionfish swallows its prey.

Lionfish and turkeyfish have a different feeding method. They swim slowly, fanning the ocean floor with their beautiful fins to uncover creatures living there. Their large fins also help the fish trap prey against rock or coral walls.

Scorpionfish use their spines for defense only, never initiating attacks on humans. When these fish are threatened, however, they erect their venomous spines and flare out the others in warning. If cornered, a scorpionfish may strike.

Mechanism of Injury

Most scorpionfish have venom glands at the base of twelve or thirteen (of eighteen) spines on their back fin, two on the belly fin, and three on the lower fin in front of the tail. The fins on the sides of these fish and on the tail do not have venom glands. When a scorpionfish strikes, venom flows into the puncture wound through grooves on both sides of

Hawai'i's turkeyfish, or *nohu pinao (Pterois sphex),* are popular in the aquarium trade. Occasionally, collectors pay a high price for these fish, which can sting a hand that strays too close. Hanauma Bay. (D. R. and T. L. Schrichte)

each spine. The spines can break off in the wound. Scorpionfish punctures most often occur on hands.

The venom of all scorpionfish is similar in composition but different in potency. The principal toxin is a heat-sensitive protein. Its main action blocks communication between nerves and muscles. Scorpionfish venom retains its full potency for twenty-four to forty-eight hours after the fish dies.

Incidence

Captive lionfish and turkeyfish sometimes sting the hands that capture them and take care of them in fish tanks.

Fishermen occasionally get stung removing a scorpionfish from a spear or hook and line, even after the fish is dead. Punctures commonly occur during the cleaning of scorpionfish.

Swimmers and divers sometimes step on or accidentally touch bottom-dwelling, well-camouflaged scorpionfish.

Several species of lionfish have caused six deaths in other parts of the world. No deaths have been reported from either of Hawai'i's two lionfish species, or any other type of scorpionfish in Hawai'i.

Only four deaths have been recorded throughout the world from the notorious stonefish. None of these were in Australia, where stonefish stings are relatively common.

Prevention

Amateur aquarists should use caution when cleaning a tank containing a lionfish or turkeyfish. The slow, graceful motions of these fish can be deceiving. Strikes are lightning fast.

Fishermen should be aware that even after a scorpionfish is dead, its venom is still powerful.

Wear thick shoes when wading in reef areas. Never reach for a submerged rock with your bare hands or feet: It could be a scorpionfish.

Signs and Symptoms

Scorpionfish venom causes exceedingly painful wounds, far out of proportion to the injury. The pain is immediate and can be savage, depending on the type of fish, the number of stings, and the age and health of the victim. Without treatment, pain peaks in sixty to ninety minutes, lasting for six to twelve hours.

Usually, except for the pain, the symptoms from a scorpionfish sting are mild. The wound and immediate surrounding area become white, then bluish. Around that, redness, swelling, and warmth to the touch may occur. A large blister occasionally appears at the sting site. Prolonged redness, swelling, and warmth signal an infection.

More serious reactions are rare in Hawai'i. After a scorpionfish sting, anxiety, headache, tremors, nausea, vomiting, diarrhea, skin rash, abdominal pain, sweating, pallor, seizures, limb paralysis, fever, and difficulty breathing signal a medically serious reaction.

Death from scorpionfish stings is extremely rare. When death does occur, it usually is in the first six to eight hours after the injury.

 # First Aid

Pain relief is a high priority in scorpionfish stings. To ease pain, soak the wound in nonscalding hot water for thirty to ninety minutes. (Victims in pain may not be able to judge if the water is too hot; someone else should test the water temperature on his or her own hand to be sure it is not scalding.) Heat inactivates at least one of the toxins in the venom and thus relieves some of the pain. If pain returns an hour or more later, try the heat soaks again.[1]

While soaking the wound, remove any protruding pieces of the spine or skin from the puncture. Do not apply ice, a tourniquet, or a pressure bandage.

To clean the wound, gently pull the edges of the skin open and scrub with gauze or a clean cloth soaked in clean, fresh water. For more information about wound care, see Part 2, *Staph, Strep,* and General Wound Care.

Victims of infected scorpionfish stings should see a doctor. If a victim shows any signs of a serious reaction to a scorpionfish puncture (listed in the Signs and Symptoms section above), regard it as a medical emergency.

Advanced Medical Treatment

No specific antivenom or clinically useful diagnostic tests exist for Hawai'i's scorpionfish stings. Australian stonefish antivenom is available at Sea World in San Diego. Reserve this antivenom for severe, systemic reactions. To date, it has not been needed in Hawai'i.

The major issue in scorpionfish puncture wounds is usually pain control. If the pain is still severe after hot water soaking, try a local anesthetic such as bupivacaine without epinephrine. Use narcotics as needed.

These wounds have a risk of retained spine fragments, which increases the severity of the envenomation and the rate of infection. Thoroughly scrub, explore, and irrigate the wound. Soft-tissue x-ray or ultrasound may help locate an embedded spine.

Occasionally, large vesicles form around the wound. Some researchers believe draining the vesicles removes venom and lessens the chances of tissue necrosis.[2] Steroids are not effective. Do not prescribe antibiotics for small lacerations with no sign of infection, except in immune-compromised patients. For more information see Part 2, Advanced Medical Treatment in *Staph, Strep,* and General Wound Care.

Because deaths have occurred worldwide from lionfish species similar to those found in Hawai'i, be prepared to support respiration and circulation. Monitor for cardiac arrhythmias. Use standard therapies for hypotension and arrhythmias.

SEA SNAKE BITES

Sea snakes, distant relatives of cobras, are often mistaken for eels. Eels, though, are fish with gills, and sea snakes are reptiles with lungs. Sea snakes are air-breathing marine reptiles. They can stay submerged for about two hours and can dive to about 300 feet.

Sea snakes eat fish, catching prey with a sideways strike of the head. The predator quickly paralyzes its prey by injecting a powerful venom through needlelike fangs, and swallows the immobilized meal whole.

The yellow-bellied sea snake is the only sea snake in Hawaiian waters, and it only rarely appears. Never touch any sea snake with your bare hand. Even dead or decapitated, the creature's persistent bite reflex can cause it to bite and envenomate. (Hal Cogger)

At least fifty-two species of sea snakes, all venomous, are found in the warm waters of the Pacific and Indian Oceans. Of these, the yellow-bellied sea snake *(Pelamis platurus)* is the world's most abundant and widespread, spanning the entire tropical and subtropical Pacific Ocean. This sea snake, the only open-ocean species, is the only one found in Hawai'i.

The yellow-bellied sea snake often floats in the marine debris of current lines, preying on the small fish that hide there. Unlike some other sea snake species, yellow-bellied sea snakes shun fresh and brackish water and therefore are not found in rivers or estuaries. Also, yellow-bellied sea snakes bear live young and do not have to come ashore to lay eggs, as some species do.

Yellow-bellied sea snakes average from 25 to 30 inches long but can grow to 44 inches. Sightings in Hawai'i are extremely rare.

Mechanism of Injury

Sea snakes have tiny, permanently erect fangs located in the front of the upper jaw. A venom duct near each fang's tip delivers venom from glands located at the rear of the head. Contrary to popular myth, sea snake teeth can, and do, penetrate human skin.

Sea snake venom is among the most potent poisons known. Beaked sea snake *(Enhydrina schistosa)* venom is the most deadly; one drop can kill three adult men. The toxicity of yellow-bellied sea snake venom is about one-fourth that of the beaked sea snake but is still potentially lethal.

Sea snake venom, which contains a mixture of proteins, is similar in all species. Because of this similarity, the same antivenom works for the bites of all sea snakes.

Sea snake venom contains nerve, heart, and muscle toxins. The extremely high potency of this venom comes from the muscle toxins, which cause muscle breakdown, kidney failure, and, occasionally, paralysis. Respiratory arrest may follow.

Incidence

No sea snake bites have been recorded in Hawai'i. Worldwide incidence is unclear, since most bites are in areas with little health care and no medical records.

The most rapid death recorded was two and one-half hours after a bite. The longest death period was twenty-four days postbite. Mortality is less than 8 percent overall, and approximately 1 percent with antivenom.

One Malaysia study showed that of 120 sea snake bites, more than 50 percent involved fishermen while they were sorting fish and handling nets.

Worldwide, only one human fatality has been attributed to a yellow-bellied sea snake. In this incident, recorded a century ago, identification of the species was questionable.

Prevention

Sea snake teeth do not penetrate most divers' neoprene wet suits but may pierce Lycra or T-shirts. Stories and photos of sport divers capturing and holding sea snakes are common, but this type of bravado is foolish. While it is true that sea snakes are not aggressive toward humans, and do not always bite, the penalty for error is severe. To be safe, never approach or touch a sea snake, living or dead. Sea snakes have a persistent bite reflex, so they can still bite, and inject venom, up to an hour after death, even after decapitation.

Most sea snake bite victims are people who handle fishing nets with sea snakes entangled in them. Before reaching a hand into a net, fishermen should check for snakes. Removal should be with a boathook or anything handy besides a bare hand. If you catch a sea snake, cut the line rather than try to remove the hook from its mouth. If you see a sea snake on a beach, do not touch it.

Report all sightings, even in nets, to wildlife authorities immediately.

Signs and Symptoms

Sea snake fangs are so small they may leave no obvious marks and often inflict little or no pain. Some victims have said they did not know they had been bitten until symptoms of poisoning began. Fang marks are one, two, or more circular dots that resemble the prick of a pin or needle. In some cases, bites look like an odd scratch mark. Serious poisoning can result from the scratch.

In about 80 percent of bites, sea snakes do not inject venom into humans. When they do, however, symptoms appear from five minutes to eight hours after the bite. Muscle weakness can cause breathing failure from a few hours to sixty hours postbite.

No one can tell from the appearance of the bite whether a snake injected venom. The site almost never shows evidence of poisoning.

A common symptom of sea snake bite victims is chills and a sense of feeling cold, whether or not venom was injected.

When envenomation does occur, early symptoms are spasm of the jaw muscles and general muscle pain on movement. These symptoms, the result of muscle damage, usually begin thirty to sixty minutes after the bite. The venom may also cause difficulty breathing, inability to speak, difficulty swallowing, excessive salivation, irregular heartbeat, muscle paralysis, muscle spasms, muscle pain, and liver and kidney failure.

Since sea snake neurotoxin does not directly affect the brain, victims who have stopped breathing will remain conscious while rescuers breathe for them.

 # First Aid

A sea snake bite is always a medical emergency, even if the victim does not appear ill. The primary goal in sea snake first aid is to get the victim to an emergency room as quickly as possible.

En route, position the bite site below the rest of the body, while keeping the victim as still as possible. Apply a broad pressure bandage over the bite about as tight as an elastic wrap to a sprained ankle. This may slow the venom's spread through the lymph system.[1] Monitor fingers or toes for pink color and warmth to be sure arterial circulation is not cut off by too much pressure.

Never cut open a sea snake bite and try to suck venom from the victim. Little is accomplished and harm can result from this old-fashioned method of snakebite treatment.

Sea snake toxin is not inactivated by changes in temperature or pH. Application of ice, hot packs, or vinegar only wastes time.

Advanced Medical Treatment

Contact the Poison Center for the location of the nearest sea snake antivenom. No clinically useful tests are available to determine sea snake envenomation. Most severe cases show symptoms within two hours of the bite. If no symptoms occur eight hours after a bite, the snake did not inject venom. While observing the patient for signs of poisoning, prepare to support breathing. Symptoms are primarily related to nerve block and muscle necrosis (described in the Signs and Symptoms section above).

Give antivenom intravenously. Use only for clinically evident systemic poisoning. Minimum dose is 1 ampoule. Normal dose is 3 to 10 ampoules, depending on the severity of symptoms. The antivenom package insert gives details about administration and risk of reaction.

Bite site infection, or necrosis, is extremely rare. No specific localized wound treatments exist. Muscle necrosis can progress to rhabdomyolysis and renal failure. Urine myoglobin is usually positive within three to six hours of the bite. Use standard rhabdomyolysis therapy, which includes monitoring electrolytes, renal function, and providing IV support. Rarely, envenomated victims may need intubation and ventilatory support.

SEA URCHIN (WANA) PUNCTURES

Sharp-spined sea urchins, known as *wana* in Hawai'i, move along the ocean floor on suction-tipped tube feet. In spite of their formidable appearance, these slow-moving, nonaggressive grazers eat mostly algae.

Hawai'i's *wana* species have two kinds of spines: long and short. The long ones are thick, hollow, and needle sharp; the shorter spines are thin, solid, and needle sharp. All are brittle and barbed.

Wana spines are permanently attached to the animal by muscles and ball-and-socket joints. The animal can wave its spines at intruders, but cannot shoot its spines like darts.

Tiny pincers, called pedicellariae, lie on the body surface of most sea urchins. The pincers can deliver a toxin that rapidly paralyzes small animals and drives predators off. The pedicellariae of Hawai'i's species are incapable of penetrating human skin.

Hawai'i has at least three species of sharp-spined sea urchins: *Diadema paucispinum, Echinothrix calamaris,* and *Echinothrix diadema.* All three species have needlelike spines, which are black or black-and-white banded. An iridescent, purple sheen may be apparent on some *wana* spines, particularly in sunlight.

Hawaiʻi has several species of sharp-spined sea urchins, or *wana*. All *wana* have needlelike spines. In Hawaiʻi, all-black *wana* are the most common. This black-and-white-banded type *(Echinothrix calamaris)*, photographed in Hanauma Bay, is less common but just as sharp. (D. R. and T. L. Schrichte)

Other kinds of sea urchins are common in Hawaiʻi. Some, like the brilliant red slate pencil, or *pūnohu (Heterocentrus mammillatus)* have harmless, pencil-shaped spines. Others, like the two rock-boring urchins, or *ʻina (Echinometra* species), have short, thick spines that, with pressure, can penetrate skin. One *ʻina* is light colored; the other is black. People sometimes mistake these dark rock-borers for small *wana*, but the two are different species.

Mechanism of Injury

Wana spines often enter the flesh when a victim steps on, falls on, or gets pushed by a wave into one or more of these animals. The barbs of the sharp-spined species may tear surrounding tissue as they enter the body.

Made of calcite and organic material, *wana* spines are brittle and break off easily in wounds. If a sea urchin spine enters a joint, severe inflammation can rapidly occur.

People have traditionally believed that *wana* carry venom on, or in, their spines. Some researchers, however, think the spines are no more toxic than any other foreign protein traumatically introduced into human tissue.

ʻIna spines create simple puncture wounds in the same manner as *wana*. The short, stout spines of these urchins sometimes break off in the flesh.

Incidence

Sea urchin punctures are common in Hawai'i. These animals are commonplace and inhabit most reefs used by board riders, snorkelers, and divers.

Serious illness from *wana* punctures is rare in Hawai'i but has been recorded here.

Prevention

Be aware that waves can push you into *wana* in shallow surge areas.

If you cannot see the bottom, do not put your feet down. The sharp spines of some sea urchins can penetrate swim fins, gloves, and beach shoes, although these may reduce the severity of the injury.

Signs and Symptoms

Punctures of spiny sea urchin spines cause an immediate, intense burning. The muscle tissue of the affected area soon begins to ache, and the skin around the puncture swells. The brittle spines usually break off deep in the flesh.

Wana contain a black or purplish pigment, which stains the skin at the wound entrance. This discoloration, called tattooing, can give a false impression that spines are lodged in the flesh.

These wounds sometimes become infected and ulcerated. *Wana*-induced nodules may appear at the surface or just under the skin two to twelve months after the injury. Rare cases of muscle aches and rash have been seen weeks after the injury, lasting several months.

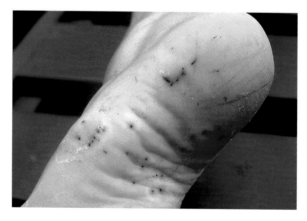

Typical sea urchin punctures, which often occur on people's feet. Small embedded spines such as these usually dissolve on their own. (World Life Research Institute photo)

Rock-boring urchins, such as this *Echinometra mathaei,* have thick spines. Another rock-boring species, *Echinometra oblonga,* is all black. The spine tips of these species are not as sharp as *wana,* but they can cause puncture wounds. Hanauma Bay. (D. R. and T. L. Schrichte)

Rarely, victims have suffered severe overall illness days or even weeks after a *wana* injury. Nausea, vomiting, numbness, muscle paralysis, abdominal pain, dizziness, low blood pressure, breathing difficulty, and paralysis have been documented in at least two Hawai'i cases. The cause of these symptoms remains unknown. The severe illnesses possibly were unrelated to the sea urchin injury.

First Aid

For simple punctures, gently pull out any protruding spines. They are so brittle they almost always break off in the wound. Neither urine nor vinegar dissolves embedded spines. Never try to crush them by hitting the area with a heavy object, which only adds to the injury. In most cases, the body either absorbs spine fragments in twenty-four hours to three weeks, or the fragments work themselves out through the skin. Most wounds heal in about one month.

Applying heat for pain control is unproven. Some doctors recommend it.[1] Others believe it is of no benefit and should not be done.[2]

The thick spines of the *'ina* do not dissolve as *wana* spines do. If *'ina* spines are embedded in the flesh (a rare occurrence) or if a sea urchin spine has penetrated a joint or nerve, see a doctor. Also see a doctor if a sea urchin wound shows any sign of infection, such as redness, warmth, or pus formation.

Victims with generalized weakness, shortness of breath, or nausea and vomiting after a puncture should go directly to an emergency room.

Advanced Medical Treatment

No antidote or specific diagnostic tests exist for sea urchin punctures.

In most patients, the primary concern in sea urchin injuries is removing the spines when possible and caring for the wound. The spines of these creatures are brittle and removal is difficult. Leave small, embedded spines in place. Over time, they usually either extrude through the skin or dissolve, leaving a purple dye that gradually dissipates. The injury caused by attempting to remove an embedded spine can exceed the damage from the spine itself.

In patients with infection or joint or nerve damage, use soft-tissue x-ray or ultrasound to find retained spines. Consider specialist referral for spines penetrating joints, nerves, vessels, or tendon sheaths.

Sea urchin puncture wounds are painful, sometimes requiring a local injection of buffered bupivacaine, or oral narcotics.

Do not prescribe antibiotics for sea urchin punctures with no sign of infection, except in immune-compromised patients. For more information about wound care, see Part 2, *Staph, Strep,* and General Wound Care.

Two systemic reactions have been reported after punctures from Hawai'i black urchins. Both victims had polyneuritis and respiratory insufficiency. One case had meningoencephalitis proven by MRI scan; the other had a variation of Guillain-Barré syndrome.[3,4] These neuritis cases may have been a reaction to a toxin, an auto-immune response triggered by embedded spines, or a coincidence. Although no proven treatment exists, search for and remove occult spines in these patients.

SHARK *(MANŌ)* BITES

About 350 species of sharks inhabit the world's oceans, most differing from one another in size, habitat, behavior, and appearance. All sharks, however, have some similarities that distinguish them from other fish. Sharks lack air-filled sacs called swim bladders that control buoyancy in most bony fish. Because of this, sharks must constantly swim or they sink to the bottom. Also, sharks have flexible cartilage rather than hard bones. Cartilage is less dense than bone, and probably aids sharks with buoyancy.

Minute, abrasive scales called denticles cover the bodies of all sharks. A brush against these sharp scales in some species is enough to abrade human skin.

Finally, all sharks reproduce through internal fertilization. Most inshore sharks of Hawai'i give birth to small numbers of live young, making them extremely vulnerable to fishing pressure.

Sharks have well-developed senses of smell and vision. They can also detect vibrations in the water, as well as electrical activity given off by the moving muscles of all living prey.

Sharks eat a wide variety of marine life, dead and alive, ranging from plankton to marine mammals. Each species has a characteristic

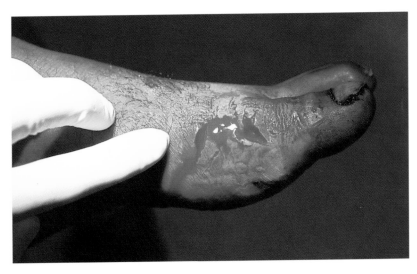

A blue shark bit the foot of this Hawai'i long-line fisherman after the fish was hauled aboard the boat. Initial treatment in Hilo consisted of cleaning the bite and giving prophylactic antibiotics. (Craig Thomas, M.D.)

The shark wound worsened after several days. X-ray revealed a piece of the shark's tooth (shown in the forceps) lodged in the wound. (Craig Thomas, M.D.)

diet. Tiger sharks *(Galeocerdo cuvier)* and great white sharks *(Carcharodon megalodon)*, the two most often implicated in fatal attacks on humans, often eat animals floating on the water's surface.

Sharks swallow food without chewing, biting large prey into pieces that fit into their mouths. Large sharks often "bite and spit," a tactic that usually mortally wounds their prey. Like the barracuda, the shark circles back to eat its kill.

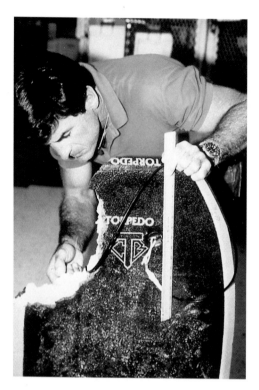

Hawai'i shark specialist John Naughton examines a boogie board with a shark bite in it. The board, found on O'ahu's North Shore, belonged to a missing surfer. (George Balazs, National Marine Fisheries Service)

At least thirty-six kinds of sharks have been recorded in Hawaiian waters, ranging from the tiny 10-inch pygmy shark *(Euprotomicrus bispinatus)* to the 59-foot whale shark *(Rhincodon typus)*. The general Hawaiian name for shark is *manō.*

Mechanism of Injury

During a bite, many sharks shake their heads and forebody, efficiently tearing flesh from the victim with razor-sharp teeth. Large sharks are known to bite with a force of approximately 20 tons per square inch, severing large bones with ease. Most fatalities come from massive tissue loss and bleeding, or panic and drowning.

Sometimes, sharks simply "bump" their victims. During these encounters, the tiny scales on sharks' skin can cause either mild or severe abrasions.

Incidence

In Hawai'i, from 1779 through February 1996, 115 incidents between sharks and humans are recorded. Of these, forty-six involved a human fatality. Ten of those deaths were regarded as the direct result of shark-inflicted wounds; ten were not regarded as deaths from sharks. The other twenty-six lacked enough details to determine the cause of death.

Of the 115 incidents, nine involved tiger sharks, two were with great whites, two were with hammerheads, and one was with a cookie-cutter shark. The rest are unknown. Attacks were reported year-round, with April having the highest incidence. Worldwide, six to ten deaths a year are from shark attacks.

Prevention

Theories abound about how to avoid shark attacks. Attacks are so rare, however, and research on sharks is so difficult, that the effectiveness of many of these strategies is unproven. Some shark behaviors are well documented. The following tactics are probably effective in preventing a shark incident:

♦ Get out of the water immediately if a tiger shark or great white shark appears anywhere in the vicinity.

♦ Gray reef sharks are territorial. If you encounter one that assumes a warning posture, arching its body and lowering its pelvic fins, back away in the direction you came and get out of the water.

♦ Never feed, chase, or try to touch or spear a shark.

♦ Trail speared fish well behind you on a long line with a float, or use a fish-keep that retains blood.

♦ Leave the water if you are bleeding.

♦ Do not swim in murky or cloudy water.

A fisherman examines a shark that was half-eaten by another shark off the Wai'anae Coast. The power of sharks' teeth and jaws is often evident on the carcasses of other fish. (John Naughton, National Marine Fisheries Service)

A local fisherman caught this 8-to-10-foot tiger shark on a baited hook just outside the reef in Waialua, O'ahu. After giving away the shark meat to local residents, the head was all that remained. (Susan Scott)

Signs and Symptoms

Skin abrasions from a shark "bump" can bleed, then ooze for hours.

Most shark bites to people are on their legs; the next most prevalent site is hands and arms. Numbness or the inability to move a finger or toe normally usually indicates tendon or nerve damage that requires repair.

Shark bites sometimes leave jagged edges. Severed arteries, severe internal injuries, and hemorrhagic shock are common results. Victims with large blood loss are often weak, dizzy, confused, pale, and sweaty. The person may be conscious but delirious. Always regard these symptoms as a medical emergency.

✚ First Aid

For skin scrapes and minor bites, scrub directly in the wound with gauze or a clean cloth soaked in clean, fresh water. Press on the area to stop bleeding. If bleeding persists or if the edges of a wound are jagged or gaping, the victim probably needs stitches. Taping a small bite shut is often an effective alternative but may leave a more visible scar than suturing. For more information about wound care, see Part 2, *Staph, Strep,* and General Wound Care.

For numbness or inability to move a finger or toe normally, see a doctor immediately.

Most shark species, such as this white-tipped reef shark *(Triaenodon obesus),* are not dangerous to humans. Divers often see these sharks resting on ledges, like this one in Hanauma Bay. (D. R. and T. L. Schrichte)

Victims who appear pale, sweaty, and nauseated are in danger of fainting. Lower the victim to the ground.

In wounds where a major artery or vein is severed, a victim can die rapidly from blood loss. Often, a rescuer can stop bleeding from large, severed blood vessels by firmly pressing anything handy (swimsuit, towel, hand) directly on the wound. Pressure usually causes the vessel to clamp down in spasm, and clots begin to form. In the water, this procedure can be nearly impossible, particularly while helping a victim to shore or to a boat. In these cases, when bleeding may be fatal, a tourniquet is appropriate. Tying a surfboard leash or dive mask strap around a massively bleeding limb could save a person's life.

Help a bleeding victim get out of the water as quickly as possible. At the beach, or in the boat, control bleeding by pressing directly on the wound, then remove any tourniquets. Leaving a tourniquet on can cause permanent injury. Maintaining pressure on the wound, take the victim to an emergency room as quickly as possible.

Advanced Medical Treatment

Shark teeth often break off inside wounds and are visible on x-ray or ultrasound. Remove the teeth meticulously to prevent infection and foreign body granuloma. Examine for tendon and nerve damage and repair, or refer, as necessary. Thoroughly scrub, explore, irrigate, and debride all shark bites. Suturing bites to control bleeding, preserve function, or improve appearance is appropriate.

Do not prescribe antibiotics for minor bites with no sign of infection, except in immune-compromised patients. Although using antibiotics to decrease the incidence of wound infection is unproven, early antibiotic therapy is reasonable for patients with large, deep lacerations. No clinical-outcome data favor a particular antibiotic regimen for preventing marine infections. Cultures of shark teeth yield *Vibrio, Klebsiella, Aeromonas,* and *Pseudomonas* species. For more information about antibiotic therapy, see Part 2, *Staph, Strep,* and General Wound Care.

For trauma victims in hemorrhagic shock, use the airway, breathing, and circulation (ABCs) protocol of trauma management.

SPONGE STINGS

Sponges come in a wide variety of colors, shapes, and sizes. Most of these primitive creatures live their entire lives in one place. None move about on their own.

A sponge's entire existence depends on water flowing through its body. The sponge creates the flow by pumping its own water. Incoming water brings the animal oxygen and food; outgoing water carries away wastes. Even sperm and eggs move in and out of the sponge on these self-made currents.

Several types of skeletons support the soft tissues of sponges. Some bear tiny, needlelike spikes, sometimes protruding through the sponge's surface. These microscopic spikes can penetrate human skin. In other sponges, tough, elastic connective tissue (called spongin) is the only support. These are the types used as bath sponges.

A stinging sponge, *Tedania ignis,* in Pearl Harbor. This species is common in Hawai'i and in parts of the Caribbean. (J. K. Sims, M.D., and Ricardo Mandojana, M.D.)

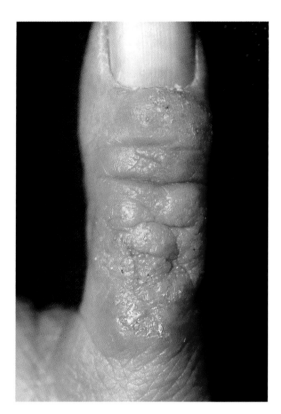

This thumb rash was caused by the sponge *Tedania ignis.* (Ricardo Mandojana, M.D.)

Hawai'i hosts at least sixty-three kinds of sponges, and twenty-four of them live only in Hawai'i. The red, yellow, or violet fire sponge *(Tedania ignis)* is one species common in Hawai'i as well as Florida and the Bahamas. This slimy-surfaced sponge is notorious for secreting a toxin causing damage to human skin.

Mechanism of Injury

Sponges produce a variety of chemical compounds interesting to researchers because of their anti-inflammatory, antibiotic, and anti-cancer properties. The negative side to these chemicals is their ability to produce skin rashes.

Two kinds of skin injuries can result from touching certain sponges. Skeletal spikes can penetrate skin, making minute puncture wounds. If sponge toxins enter these wounds, this usually minor injury is more severe.

The other possible sponge injury is a reaction to a chemical substance found on the surface of some sponges, such as Hawai'i's fire sponge. These reactions range from mild rash to severe generalized illness.

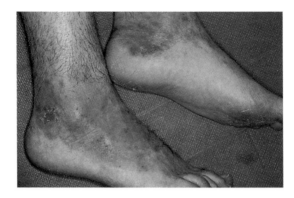

This foot rash was caused by the sponge *Tedania ignis.* (Ricardo Mandojana, M.D.)

Incidence

Divers, snorkelers, swimmers, and anglers are at risk for sponge stings. Sponges sometimes get caught in nets or on hooks. Most sponge stings are on the hands, but bare legs can be stung by brushing a fire sponge in passing, or by getting pushed into one by surge.

Prevention

Not all sponges cause skin injuries, but to be safe, do not touch any. Divers should wear protective clothing, including gloves. Beware of beached sponges; dried sponges can remain toxic.

Signs and Symptoms

Most sponge injuries appear as mild to moderate rashes that itch and burn after contact. This sometimes progresses to swelling, blisters, and stiffness, particularly if the sting is near a joint. Affected skin is often red and mottled. Blisters may drain and become infected. In some cases, the skin peels ten days to two months after contact.

If the sponge touches a large area of skin, the victim may have fever, chills, dizziness, nausea, and muscle cramps. An allergic reaction to a severe sponge sting can develop a week to ten days after a sting.

✚ First Aid

For uncomplicated sponge stings, rinse and dry the area. Lift skeletal spikes from the skin with sticky tape. You may not be able to see the tiny cactuslike spikes.

Other treatments for sponge stings are unproven. In some cases, soaking the area in vinegar or rubbing alcohol seems to relieve pain.[1] After soaking, try 1 percent hydrocortisone ointment four times a day, and one or two 25-milligram diphenhydramine (Benadryl) tablets every six hours. These drugs are sold without prescription. Diphenhydramine may cause drowsiness: Do not drive, swim, or surf after taking this medication.

Tetanus-causing (lockjaw) bacteria live in some sponges. Make sure your tetanus immunization is up to date. For information on tetanus immunization, see Part 2, *Staph, Strep,* and General Wound Care.

Infections can develop after sponge stings. If blisters fill with pus or the red area spreads and feels warm, see a doctor.

Any difficulty breathing or a generalized body rash after a sponge sting is always a medical emergency. (See Part 2, Allergic Reactions.)

Advanced Medical Treatment

No specific antidote or clinically useful diagnostic tests exist for sponge stings. Most stings have only localized symptoms that usually respond to first aid treatment.

Some victims suffer generalized allergic reactions. These patients require standard therapy of epinephrine, antihistamines, rehydration, and airway support. See Allergic Reactions.

For erythema multiforme, try oral steroids for several days. For information about secondary infections, see Part 2, *Staph, Strep,* and General Wound Care.

SQUIRRELFISH *('ALA'IHI)* STINGS

Most species of squirrelfish, or *'ala'ihi* (family Holocentridae), are shy, deep-water fish, seldom seen by snorkelers. These red, big-eyed fish take shelter during the day in caves and under ledges, venturing out at night to search for crabs and shrimp on the ocean floor. Hawai'i hosts several species of squirrelfish. All have a spine, sometimes venomous, on the lower corner of each gill cover.

Some squirrelfish are good to eat and are sought by anglers. The longjaw squirrelfish, *Sargocentron spiniferum,* grows up to 18 inches long and is the largest of its family. It is notorious among anglers for its painful sting.

The longjaw squirrelfish *(Sargocentron spiniferum)* is the largest squirrelfish in Hawai'i, growing to 18 inches. These squirrelfish are notorious among Hawai'i anglers for the painful sting caused by the spine on the fish's gill covers. (John Hoover)

Mechanism of Injury

Some, perhaps all, squirrelfish bear a venomous spine on the lower part of each gill cover. A captured fish thrashing at the end of a spear, line, or on the deck of a boat can drive one of the spines into an angler's hand or foot, usually causing a puncture wound.

Pain from these wounds is probably the result of tissue trauma, the effects of the spine's venom, and the introduction of slime and other surrounding substances into the wound.

Prevention

Anglers should wear gloves and boots while removing live fish from spears or hooks. While cleaning squirrelfish, particularly the longjaw squirrelfish, keep your fingers away from gill cover spines.

Incidence

No data are available on the frequency of squirrelfish wounds in Hawai'i or elsewhere. Hawai'i anglers have reported painful wounds from squirrelfish stings.

Judging from the severe pain they cause in human skin punctures, squirrelfish gill spines are considered venomous. The offending spine is visible in this photo. (John Hoover)

Signs and Symptoms

Longjaw squirrelfish punctures cause immediate pain far out of proportion to the injury. One angler reported a sting from this fish that was so painful he nearly fainted. Swelling and bleeding may also occur.

✚ First Aid

For pain relief, try soaking the area in hot, nonscalding water. This method of easing pain has not been studied with squirrelfish wounds, but works for other fish stings, such as rays and scorpionfish. (Victims in pain may not be able to judge if the water is too hot; someone else should test the water temperature on his or her own hand to be sure it is not scalding.)

For minor squirrelfish wounds, gently pull the edges of the skin open and remove any embedded material either by rinsing or using tweezers. Then scrub directly inside the cut with gauze or a clean cloth soaked in clean, fresh water. Press on the wound to stop bleeding. If bleeding persists or if the edges of a wound are jagged or gaping, the victim probably needs stitches. Taping a cut shut is often an effective alternative but may leave a more visible scar than suturing. For more information about wound care, see Part 2, *Staph, Strep,* and General Wound Care.

Victims with any feeling of overall illness after a squirrelfish puncture should go directly to an emergency room.

Advanced Medical Treatment

No antivenom or diagnostic tests exist for this type of envenomation. No studies exist on squirrelfish venom.

For most patients, pain control, foreign body detection and removal, and localized wound management are the primary issues. Treat pain unrelieved by heat with buffered bupivacaine. Use soft-tissue x-ray or ultrasound to locate pieces of spine that may have broken off in the wound. Do not prescribe antibiotics for minor wounds with no sign of infection, except in immune-compromised patients. (See Part 2, Advanced Medical Treatment in *Staph, Strep,* and General Wound Care.)

STINGING LIMU STINGS

People in Hawai'i call a variety of marine organisms stinging limu (seaweed), but only one true seaweed is known to commonly cause a rash in humans here. This blue-green seaweed, called *Lyngbya majuscula* (also called *Microcoleus lyngbyaceus* in some reports), usually grows in clumps and looks like dark, matted masses of hair or felt. This seaweed is usually blackish green or olive green, but it also grows in shades of gray, red, or yellow.

The filaments of *Lyngbya majuscula* grow from 2 to 4 inches long. They often tangle with other seaweeds on reef flats, in tide pools, or in deeper water up to 100 feet. When loose in the water, this seaweed looks like floating, tangled strands.

Mechanism of Injury

The toxicity of this seaweed varies greatly depending on region, season, and type. Not all strains are toxic. No one knows what causes this variation.

When toxic, *Lyngbya majuscula* contains two potent, inflammatory toxins; both cause skin damage on contact. Seaweed fragments typically get caught inside swimming suits, rubbing the toxins into the skin.

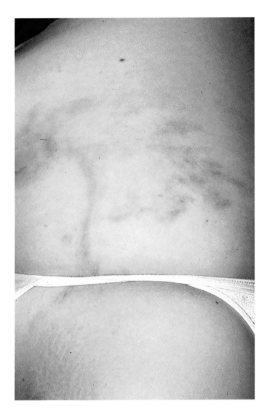

This woman's abdomen shows the classic red, blistered rash caused by *Lyngbya majuscula,* or stinging limu. Wai'anae Coast. (Scott Norton, M.D.)

Rarely, strong onshore winds create ocean spray that can blow seaweed fragments into the faces and eyes of people walking along a beach, causing symptoms. Inhaling fragments of the seaweed may cause severe lung inflammation.

Chewing on this seaweed burns the mouth and lips. Swallowing it causes fatalities in laboratory mice; it could be fatal in humans.

People in Hawai'i do not intentionally eat this type of seaweed.

Incidence

Epidemics of this seaweed-induced rash occasionally occur in Hawai'i and in Okinawa. In Hawai'i, the highest number of cases are from June through September, when toxic fragments drift into swimming bays and beaches. In Okinawa, people walking on beaches where winds carry ocean spray have suffered face rash and eye irritation.

After a storm in Lahaina, one person died from severe lung inflammation, perhaps from inhaling fragments of this seaweed. Another person on the island of Hawai'i accidentally tried to eat some of this toxic seaweed, suffering burns of the mouth.

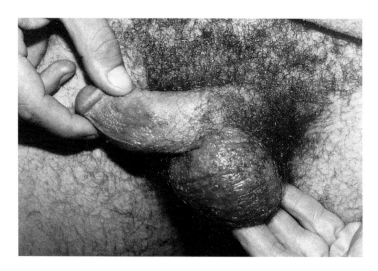

Lyngbya majuscula (also known as *Microcoleus lyngbyaceus*), or stinging limu, from Windward Oʻahu caused this painful rash. Such injuries are typical of this seaweed. Small pieces of it drift into people's swimsuits, often washing down to the genital and anal regions, where the seaweed bits lodge and rub into tender skin. (D. Huntley, M.D.)

Prevention

The Hawaii Department of Health issues public warnings when outbreaks of this rash are reported by swimmers. Heed the warnings. Common areas are Kāneʻohe Bay, Kailua Bay, and the waters off Lāʻie, but leeward waters are also involved.

Avoid swimming through masses of seaweed.

Limiting swimming time in affected areas does not guarantee protection, nor does prompt removal of a bathing suit with immediate showering after swimming. Exposure for any length of time causes a rash in some people.

Eat seaweed harvested only by experienced limu pickers.

Signs and Symptoms

Rash victims feel an itching and burning sensation within minutes or up to twenty-four hours after leaving the water. A red, sometimes blistering rash develops, typically in a swimming suit pattern. Often, the rash is more intense around the genital and anal area, where the seaweed gets trapped and is rubbed in by the suit. Prolonged contact produces blisters that look like burns, producing constant pain.

The rash can also develop on the face, and in the eyes and mouth. Some victims have swelling of the eyes and mouth, but no rash.

Eating this seaweed produces almost immediate, intense burning of the lips and mouth, causing the victim to spit it out. Pain lasts for about three days, but the mouth does not return to normal for about two weeks. No cases of systemic poisoning have been reported, but swallowing *Lyngbya majuscula* is potentially fatal.

First Aid

For mild to moderate cases of skin rash, remove the swimming suit immediately and wash the skin vigorously with soap and water. Wash the suit too.

Although unproven, cool compresses or rubbing alcohol may help relieve the pain. For persistent itching or skin rash, try 1 percent hydrocortisone ointment four times a day, and one or two 25-milligram diphenhydramine (Benadryl) tablets every six hours. These drugs are sold without prescription. Diphenhydramine may cause drowsiness: Do not drive, swim, or surf after taking this medication.

Irrigate eye stings with tap water for at least fifteen minutes, and then go to a doctor. For severe discomfort, blistering that does not respond to first aid treatment, or any sign of infection (pus, redness, swelling), also see a doctor.

Victims who have accidentally eaten this limu should suck on ice to relieve pain while going to an emergency room.

Any difficulty breathing signals an allergic reaction, which is always a medical emergency. (See Part 2, Allergic Reactions.)

Advanced Medical Treatment

No specific antidote or clinically useful diagnostic tests exist for stinging limu. No known treatment for ingestion of this seaweed exists. Local treatment is similar to burn care.

Secondarily infected stings may need debriding and antibiotics. *Vibrio* organisms have been cultured from stinging limu. (See Part 2, *Staph, Strep,* and General Wound Care and *Vibrio* Infections.)

For lip, mouth, and throat burns, sucking on ice offers the best pain relief. Neither diphenhydramine nor viscous lidocaine relieves oral symptoms. One patient treated with steroids was symptom free after three days and completely healed in two weeks.[1]

Multiple other marine organisms, including tiny jellyfish and flatworm larvae, can cause similar-looking skin rashes. Distinguishing these from limu rash is often impossible. Treatment is the same.[2] The potent topical steroid clobetasol may worsen symptoms, which usually last ten to twelve days.[3]

SURGEONFISH CUTS

Hawai'i hosts twenty-three species of surgeonfish (family Acanthuridae), also called tangs or doctorfish. Most surgeonfish are herbivores, but a few eat animal plankton. Some types of surgeonfish are territorial loners, but others cruise the reef in large schools. Adults range in size from about 5 to about 30 inches long.

Surgeonfish get their name from the sharp spine, or scalpel, the fish bear immediately in front of the tail fin on each side. Some species can expose this spine, normally tucked flat into a groove, to fend off intruders or predators. In other species, the spines are fixed erect. The area around the scalpels, or the scalpels themselves, are often brightly colored, presumably a warning of their potential danger. Young surgeonfish are transparent, with venomous dorsal (back), pelvic (belly), and anal (rear belly) spines.

Unicornfish, or *kala* (*Naso* species), are also members of the surgeonfish family. *Kala* bear two scalpel-like spines on each side, one behind the other. The spines of some adult unicorn fish may be venomous.

Surgeonfish are active during the day, seeking shelter on the reef at night. To many people, these fish are good eating.

The yellow tang, or *lau'īpala (Zebrasoma flavescens)*, is also called a surgeonfish because of the scalpel-sharp spines it carries in front of the tail. A white mark outlines these spines on yellow tangs. Hanauma Bay. (D. R. and T. L. Schrichte)

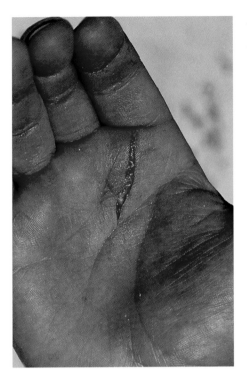

A surgeonfish cut this person's hand during a session of fish feeding. (B. Letot, M.D.)

Mechanism of Injury

A swipe of a surgeonfish spine usually produces a deep cut or, sometimes, a puncture wound. The pain occasionally is worse than the wound warrants, suggesting envenomation in some species.

Incidence

Surgeonfish can accidentally slash divers or snorkelers during feeding sessions. Anglers sometimes get cut while removing surgeonfish from a net, spear, or fishhook. When cornered, some surgeonfish species lash out at divers.

Prevention

If you feed fish while snorkeling or diving, toss the food away from yourself and others.

Wear heavy gloves when handling surgeonfish in nets or on spears.

While diving, do not chase or corner surgeonfish, especially loners such as the Achilles tang, or *pāku'iku'i (Acanthurus achilles)*. This territorial fish is aggressive, even to its own kind.

Signs and Symptoms

Surgeonfish cuts are usually short and deep, and bleed freely. If the spine is venomous, pain is moderate to severe, with a burning sensation. Localized muscle aches and swelling can develop around the cut.

Numbness or the inability to move a finger or toe normally may indicate tendon or nerve damage that requires repair.

Overall illness from these cuts is rare.

 # First Aid

For minor cuts, gently pull the edges of the skin open and remove any embedded material either by rinsing or using tweezers. Then scrub directly inside the cut with gauze or a clean cloth soaked in clean, fresh water. Press on the wound to stop bleeding. If bleeding persists or if the edges of the wound are jagged or gaping, the victim probably needs stitches. Taping a cut shut is often an appropriate alternative but may leave a more visible scar than suturing. For more information about wound care, see Part 2, *Staph, Strep,* and General Wound Care.

No studies exist on treating surgeonfish cuts, but hot water soaks relieve pain in ray and scorpionfish stings. For pain control, try soaking the wound in nonscalding hot water for thirty to ninety minutes. (Victims in pain may not be able to judge if the water is too hot; someone else should test the water temperature on his or her own hand to be sure it is not scalding.)

If pain worsens, redness and swelling increase, or a finger or toe will not move normally after a surgeonfish cut, see a doctor.

 # Advanced Medical Treatment

Documentation on surgeonfish spine toxin is sparse, but the toxin appears to cause only localized pain and swelling. Spines can break off inside wounds and are visible on soft-tissue x-ray. Ultrasound may also help locate pieces of surgeonfish spine. Remove them meticulously to prevent infection and foreign body granuloma.

Thoroughly scrub, explore, irrigate, and debride all wounds. Examine for tendon and nerve damage and repair, or refer, as necessary. Suturing wounds to control bleeding, preserve function, or improve appearance is appropriate.

Do not prescribe antibiotics for minor injuries with no sign of infection, except in immune-compromised patients. For more information about antibiotic therapy, see Part 2, *Staph, Strep,* and General Wound Care.

Part 2

Infections and Poisonings

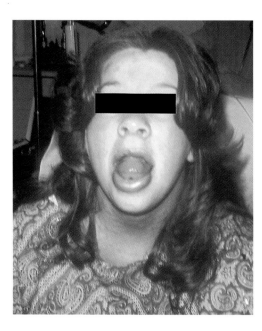

Something this woman ate caused her tongue to swell so large that she could not close her mouth. Swelling like this is typical of severe allergic reactions and is dangerous because it can block air flow through the breathing passages. After an emergency room doctor treated this allergy with epinephrine and antihistamines, the swelling receded and the patient went home. (Jerry Hughes, M.D.)

ALLERGIC REACTIONS

When a person's immune system perceives a foreign substance to be harmful, and most other people are not particularly sensitive to that substance, the person is said to be allergic or hypersensitive. For example, skin redness and welts after a person brushes against a coral head is not an allergic reaction because most people have similar symptoms. Some people, however, are hypersensitive to living coral. If a minor coral touch causes blistering, severe pain, or overall illness, the person is probably allergic to something in coral.

Anyone can be allergic to any plant or animal, even if they have never had a previous allergy. Sometimes, allergies develop through frequent exposure to a substance. In other cases, an allergic reaction occurs after only one or two exposures.

Allergies to marine substances can be mild, moderate, or life threatening.

Mechanism of Injury

An allergy to a foreign substance prompts specialized cells in the body to release histamine and other chemical "messengers." These messengers cause tiny blood vessels to leak fluids into the surrounding tissues, which makes the area red and swollen.

When the body responds to a substance by releasing massive amounts of histamine and other chemical messengers, blood vessels leak throughout the entire body. Blood pressure falls and breathing passages swell.

For an allergic reaction, a person must have eaten or touched the plant or animal at least once, and perhaps many times, before the attack. Once a person has become sensitized to a substance, even minimal contact can set off a serious reaction. In allergies, reactions are usually more severe with each subsequent exposure.

Incidence

Mild to moderate allergies to marine substances are fairly common but usually pose no threat to life. Severe allergies involving the entire body are rare but can occur without warning.

Shellfish allergies are one of the most common food allergies in the United States. Fish allergies also occur, but some of them turn out to be scombroid poisoning, which mimics fish allergy. Because of this, some people falsely believe they are allergic to fish. (See Scombroid Fish Poisoning.)

Some people are allergic to substances on the skin of certain fish.

Prevention

To prevent allergic reactions, stay away from any plant or animal that has caused you a previous allergic reaction. A minor reaction the first time may mean a major reaction the next.

Even if a food or substance has not caused a problem in the past, a person may still develop an allergy to it. Allergies like this persist through the person's lifetime.

Signs and Symptoms

A skin response usually shows up as a red, raised welt (hive) on the affected area, sometimes blistering. Itching or pain in the area is common.

Symptoms of moderate allergy are red and running eyes and nose, swelling of the lips and tongue, dizziness, nausea, vomiting, hives, and diarrhea.

A hoarse voice, a sensation of thickening in the throat, or difficulty drawing a deep breath are signs of severe, life-threatening allergy.

Many allergic reactions begin within fifteen to thirty minutes of contact. Nearly all develop within six hours. Allergy can be delayed, however, when toxins or foreign substances gradually dissolve into body tissue.

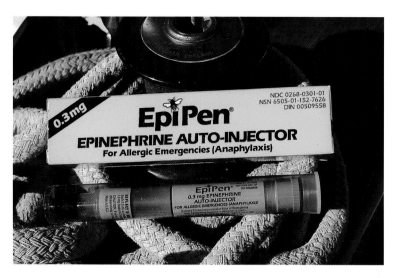

Dive leaders, fishermen, and anyone who has had a previous serious allergic reaction should carry one of these epinephrine (adrenaline) injectors when off-shore or far from medical help. The prescribing doctor should explain when and how to use this emergency drug to treat breathing difficulties caused by allergy. (Susan Scott)

 # First Aid

For mild or moderate allergic reactions, try one or two 25-milligram diphenhydramine (Benadryl) tablets every six hours. This drug causes drowsiness: Do not surf, swim, or drive after taking it. Apply 1 percent hydrocortisone ointment to hives for itching. The drugs are sold without a prescription.

If the allergy is from surface contact, wash the affected area thoroughly. Change out of contaminated clothing and wash it.

Severe allergies can cause life-threatening breathing symptoms. If a victim of any marine wound or seafood allergy develops swallowing or breathing difficulties of any kind, rush the victim to the nearest emergency room.

Epinephrine (adrenaline), the emergency treatment for breathing distress caused by allergy, can be life saving in areas remote from hospitals.[1] This injectable medication comes in an auto-injector, available by prescription. Dive leaders and people who have had a previous severe allergic reaction should carry one on all ocean outings. Administer the epinephrine as soon as an allergy victim begins wheezing or complains of tightness in the chest.

Epinephrine is a potent medication causing rapid heartbeat, anxiety, and high blood pressure. Rarely, it can cause heart attacks or strokes. Reserve this medication for victims with breathing problems.

Advanced Medical Treatment

Thoroughly wash allergens from the affected area. Antihistamines are effective initial treatment for allergic reactions. Hydroxyzine is slightly more effective than diphenhydramine. For a persistent rash resistant to standard antihistamines, use cimetidine.

Epinephrine relieves hives, itching, and respiratory symptoms. For patients with bronchospasm, inhaled therapy with a beta-agonist, such as albuterol, is often effective. Oral steroids may decrease the duration of symptoms.

For severely allergic patients who have eaten fish or shellfish within one hour, consider gastric lavage followed by charcoal.

Rarely, allergy victims develop systemic shock requiring intravenous saline for hypotension. For severe hypotension or bronchospasm not responding to initial therapy, try epinephrine, 0.1 milligram intravenously, and intravenous steroids.

BOTULISM (FROM FRESH FISH)

The bacterium that causes botulism, *Clostridium botulinum,* is common in soil, mud, sand, and sea-bottom silt. This infamous organism is a strict anaerobe, meaning it cannot live in the presence of oxygen. When conditions are right for *Clostridium botulinum,* the organism produces an exotoxin, a powerful poisonous protein released into the area around the organism. In humans, this exotoxin causes botulism.

The eye-stripe surgeonfish, or *palani,* is the only fresh fish ever implicated in fish botulism in the United States. Hanauma Bay. (Peter Dunn-Rankin)

Clostridium botulinum reproduces by spores. The spores are not usually killed by heat, cold, light, drying, or salting. These methods, however, can prevent the spores from producing the potent botulism exotoxin. Once produced, the exotoxin can be destroyed by heat.

Mechanism of Injury

This bacterium's exotoxin is the most potent biological toxin known to humans. Approximately one-sixtieth of an ounce (0.5 gram) is a lethal dose. The toxin blocks communication between nerves and muscles, thus creating life-threatening paralysis.

Botulism can result from a person eating preformed toxin from an improperly preserved fish.

Rarely, spores infect a wound or grow in the human digestive tract, producing the deadly exotoxin.

Incidence

In 1990, botulism struck three people from a Maui family after they ate an eye-stripe surgeonfish, or *palani (Acanthurus dussumieri),* the day they bought it. The victims bought the fish at a market where the cooling equipment was broken. Remnants of intestine remained inside the inadequately cleaned fish, which a fisherman had sold to the market four to fifteen days before the purchase.

In this incident, two of the three victims ate the fish's intestines. The other victim touched the intestines and then ate the fish's head. This suggests that without proper cooling, fresh-fish intestines may provide an ideal environment for *Clostridium botulinum* to grow and produce exotoxin.

This is the only case of botulism from fresh fish ever reported in the United States. All three victims survived.

Healthy adults who eat the spores of this organism almost never get botulism poisoning, because stomach acid prevents production of exotoxin. Only about thirty cases of adult botulism are reported in the United States each year. Eating improperly preserved fish causes about 13 percent of food-borne botulism cases, mainly in Alaska.

Prevention

Botulism can be prevented four ways: (1) By destroying the bacterium's spores, (2) by preventing the spores from producing exotoxin, (3) by destroying the exotoxin, and (4) by avoiding high-risk foods.

Since it is extremely difficult to destroy the spores of most strains of *Clostridium botulinum,* the most practical prevention is to keep these hardy spores from producing exotoxin. To do this, keep fresh fish cold at all times before cooking. Freezing, drying, and salting fish also discourages exotoxin growth.

To destroy any botulism exotoxin that may have grown in a fish, boil the fish for ten minutes or heat it to about 180 degrees F (80 degrees C) for thirty minutes.

Signs and Symptoms

Within six to seventy-two hours after eating fish containing *Clostridium botulinum* exotoxin, victims usually experience nausea, vomiting, abdominal pain, and diarrhea. These symptoms progress to dry mouth, hoarseness, difficulty swallowing, facial weakness, drooping eyelids, nonreactive or sluggishly reactive pupils, dilated pupils, blurred or double vision, muscular weakness leading to paralysis, and respiratory paralysis. With breathing support, victims usually remain conscious and alert.

Death occurs in 10 to 50 percent of cases.

First Aid

Botulism poisoning is a medical emergency. Victims can stop breathing with little warning. If this happens, start cardiopulmonary resuscitation (CPR) and rush the victim to the nearest emergency room. Save suspect food for laboratory analysis.

Advanced Medical Treatment

Make the initial diagnosis by history and physical exam. Obtain specimens of blood, stool, and any fish remains for toxicology and culture. Do not delay treatment by waiting for toxicology results. Handle all specimens with gloves.

Suspect botulism in patients with descending paralysis. These patients can develop respiratory arrest rapidly. Be prepared to intubate. Arterial blood gases are a poor indication of impending respiratory arrest. Use pulmonary function tests. If the patient's function is less than 30 percent of predicted values, respiratory arrest is imminent.[1]

The primary differential diagnosis is Guillain-Barré syndrome, usually an ascending paralysis. Another possibility is myasthenia gravis. A five-minute response to Tensilon test suggests this disease.

Clostridium botulinum exotoxin is among the most dangerous toxins in the world, irreversibly binding to the neuromuscular junction. A specific antidote called *botulinum* antitoxin, made from horse serum, may prevent this binding and thus prevent respiratory failure. *Botulimum* antitoxin is available from the Hawai'i branch of the U.S. Public Health Service. The decision to administer this drug must be in consultation with the Hawaii Department of Health. Before administration, give an IV test dose for sensitivity. Treat as necessary with antihistamines, steroids, and epinephrine.

Botulism bacteria can infect wounds, producing exotoxin that spreads throughout the body. These infected wounds, which may have no cellulitis, require thorough debridement. Antibiotics are not effective in prevention or treatment of botulism wound infection. *Clostridium botulinum*–infected wounds are at high risk for tetanus.

Report all cases of botulism to the Hawaii Department of Health.

CHOLERA

Cholera is a severe diarrheal illness caused by a bacteria called *Vibrio cholerae*. This organism lives in warm seawater and estuaries, usually in areas where sanitation is poor. The disease is often associated with eating shellfish.

In small numbers, cholera bacteria do not usually cause illness in people. But because shellfish, such as oysters and clams, eat by filtering water, bacteria can become concentrated in their bodies. Apparently, build-ups of cholera organisms do not harm the shellfish. They do, however, cause illness when humans eat the shellfish raw or partially cooked.

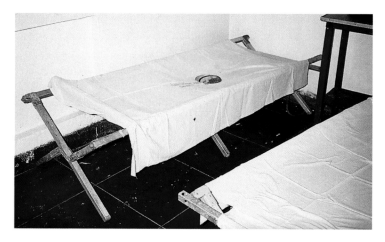

This cholera bed is in a Manila hospital. Workers place a bucket under the hole in the cot. In cholera, fluid loss from diarrhea can be up to 1 quart an hour. (Vernon Ansdell, M.D.)

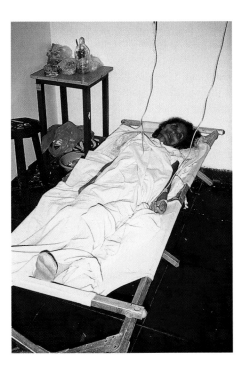

This cholera patient in Manila, the Philippines, is vastly improved from the day before. She was treated with oral rehydration solutions and IV fluids. (Vernon Ansdell, M.D.)

Another way cholera bacteria can multiply to dangerous levels is by improper refrigeration of fish or shellfish that contain the organism.

Other *Vibrio* bacteria, including some strains of *Vibrio cholerae,* can cause diarrheal illnesses less severe than cholera. These bacteria can also cause overall illness by spreading throughout the body via the bloodstream, and from wound infections. (See *Vibrio* Infections.)

Mechanism of Injury

Cholera arises from eating raw or undercooked contaminated seafood, from drinking contaminated water, or from eating undercooked food prepared with contaminated water.

In healthy people, infection requires at least one million *Vibrio cholerae* bacteria. Once the bacteria are in the body, they multiply in the small intestine, producing a toxin. The toxin disrupts the balance of water absorption and secretion in the intestine, producing severe diarrhea and rapid dehydration. Because the dose required to infect a person is large, cholera rarely spreads from person to person.

Incidence

In Hawai'i, at least four cases of cholera, in two episodes, have been reported in the 1990s. These cases probably came from eating imported fish infected with *Vibrio cholerae,* although the actual source was never found. All the patients survived.

Cholera has been the culprit in seven deadly worldwide outbreaks since the early 1800s. The latest began in Indonesia in 1961 and continues to spread throughout the world. In 1973, cholera appeared in the United States after being absent since 1911. New United States cases continue to appear.

People who have less stomach acid than normal have a higher risk of developing cholera when exposed to the bacteria. Decreased acid can be the result of stomach surgery, or it can come from taking certain ulcer medications, including antacids. People with damaged immune systems are also at higher risk.

Prevention

Cold temperatures prevent cholera bacteria from multiplying to dangerous levels; hot temperatures kill them. Therefore, keep raw shellfish refrigerated until ready to cook, and cook it thoroughly. Refrigerate leftovers in case any bacteria still live.

People who have had stomach surgery or take ulcer medicines should not eat raw (or undercooked) oysters, clams, crabs, or other shellfish.

People with immune-system damage, such as those with diabetes, liver disease, splenectomy, AIDS, valvular heart disease, or anyone taking steroids, should not eat raw or undercooked shellfish. They risk blood-borne infections.

Signs and Symptoms

Sometimes people can be infected with cholera and have no symptoms. Sixty to 80 percent of cases are so mild, cholera is not suspected.

In moderate to severe cases, diarrhea begins one to five days after eating or drinking the bacteria. Serious cases produce watery diarrhea (often described as "rice water") at approximately 1 quart per hour. This can rapidly lead to severe dehydration and death. Vomiting may or may not occur. Fever, if present, is low grade.

With medical treatment, cholera usually lasts three to five days. Peak water loss occurs about twenty-four hours after the illness begins.

 # First Aid

Treat shellfish-induced diarrhea with sports drinks such as Gatoraid. The drinks contain salts and sugar similar to fluids lost in diarrhea. To make an effective drink, add 1/2 teaspoon table salt and 2 tablespoons sugar to a quart of water. Intake should at least equal the amount of fluid lost. Most people are cured with this therapy. A few people need intravenous fluids.

Sometimes, bacteria become blood-borne, causing life-threatening illness. See a doctor for severe, ongoing diarrhea, fever, rash, or generalized illness.

Advanced Medical Treatment

Untreated, severe cholera has a mortality rate of 50 percent. Treated, the rate is less than 1 percent. Replace fluids and electrolytes aggressively. Oral rehydration solutions successfully treat 90 percent of cases not in shock. Do not wait for laboratory results to rehydrate a victim.

Doxycycline and fluoroquinolones, such as ciprofloxacin, shorten the duration of the diarrhea and reduce fluid loss. Antibiotics, however, are not as important as fluid therapy. Treatment for enteritis from non-cholera *Vibrio* is similar. These patients are at risk for invasive disease and sepsis. (See *Vibrio* Infections.)

Make diagnosis with darkfield microscopy or stool culture. Report all suspected cholera cases to the Hawaii Department of Health immediately.

CIGUATERA FISH POISONING

Ciguatera is a poisoning people usually get from eating reef fish. The primary culprit in this poisoning is a single-cell marine organism, *Gambierdiscus toxicus.* This naturally occurring organism sometimes blooms among seaweeds. When plant-eating reef fish graze on the seaweeds, they also swallow the ciguatera-causing *Gambierdiscus toxicus.*

The toxins produced by *Gambierdiscus toxicus* do not seem to harm the fish, but remain in their flesh and organs. When larger fish eat the grazers, the toxin transfers to the predators' flesh and organs, again with no apparent harm to the fish. People are not so fortunate. When humans eat fish containing the toxins of *Gambierdiscus toxicus,* serious illness often results.

Mechanism of Injury

In the human body, ciguatoxin alters the way sodium moves into nerve and muscle cells. This alteration affects a wide range of bodily functions. (See Signs and Symptoms, below.)

The goldring surgeonfish, or *kole (Ctenochaetus strigosus),* is often implicated in ciguatera poisoning. Hanauma Bay. (D. R. and T. L. Schrichte)

Incidence

Ciguatera is the most common cause of seafood poisoning in the United States. Since this toxin is associated with coral reefs, Hawai'i and Florida have the greatest incidence. Hawai'i averages about eighty reported cases a year. Many cases are mild and the symptoms are sometimes vague, so the actual number is higher.

Ciguatera is also common in certain Caribbean and South Pacific countries. Approximately fifty thousand cases a year are recorded worldwide.

Fatalities in ciguatoxin poisoning are rare. In a study of 12,890 cases, twenty-two people died. Some researchers believe the fatalities are actually caused by either palytoxin (see Zoanthids) or clupeotoxin (see Sardine Poisoning). Three deaths from ciguatoxin poisoning have been reported in Hawai'i.

Some people believe that the incidence of ciguatera fish poisoning is increasing in Hawai'i and throughout the world because of natural and human-induced alterations of the shoreline and ocean floor. It is well documented in the South Pacific that *Gambierdiscus toxicus* growth increases after hurricanes. The effects of dredging and water pollution on the growth of *G. toxicus* need more study to draw conclusions for Hawai'i.

It is true that more cases of ciguatera have been reported in Hawai'i (and worldwide) during the past two decades than ever before. More people are fishing and eating fish than ever before. Also, public awareness of this poisoning is higher today than in previous decades, resulting in more reports of the illness.

The number of cases reported in Hawai'i over the past six years shows no trend:

1989	84
1990	52
1991	138
1992	34
1993	86
1994	63
1995	51

No cases of ciguatera have been reported from Moloka'i or Lāna'i. Certain areas of the other main Hawaiian Islands are known for producing a high number of ciguatoxic fish, but the areas change over time. Most cases come from recreational catches rather than from fish bought in markets.

All species of reef fish should be suspect, including reef sharks. In Hawai'i, the three highest-risk fish for ciguatera are jacks, also called *ulua, pāpio,* and *kāhala* (Carangidae); gold-ring surgeonfish, or *kole (Ctenochaetus strigosus);* and forktail snapper, or *wahanui (Aphareus furca).*

One case of ciguatera has been reported in a child who ate a jellyfish (species unknown) from American Samoa. In Hawai'i, one case of ciguatera resulted from eating an octopus; another from a skipjack tuna, or *aku (Katsuwonus pelamis).* No other open ocean fish, such as *mahimahi* or *ono,* appear to carry ciguatoxin.

In Australia, two people reported that local ciguatera symptoms were transmitted sexually, one male to female, one female to male.

Prevention

Despite popular folklore, there is no way to identify a ciguatoxic fish by look, color, smell, taste, or texture. The toxins causing ciguatera poisoning are neither destroyed nor inactivated by cooking, canning, drying, freezing, or smoking.

To minimize the risk of eating a ciguatoxic fish:

♦ Do not eat reef fish or fish that feed on reef fish.

♦ If you do eat reef fish, do not eat the fish's internal organs, head, or eggs. Ciguatoxin can be one hundred times more concentrated in these parts of the fish. Also, eat only a small portion of a suspect fish. Severity of illness is directly related to the amount of toxin eaten.

  ♦ Know that all inshore-feeding fish, including sharks and bar-
    racuda, are suspect.

  ♦ Remember which species are most likely to carry ciguatoxin.
    Because the toxin accumulates in fish tissues, the bigger the
    fish, the more toxic it may be.

  ♦ Although it is rare, be aware that jellyfish, octopus, and perhaps
    other invertebrates may also carry ciguatoxin.

Ciguatera test kits currently are not available to the public.

Signs and Symptoms

The normal passage of sodium in and out of nerve and muscle cells reg-
ulates a wide range of bodily functions, so ciguatoxin poisoning has
multiple effects. One hundred seventy-five symptoms have been
reported from ciguatera poisoning.

Symptoms begin ten minutes to thirty-six hours after eating a toxic
fish. Nausea, vomiting, and diarrhea are usually followed by general
weakness and muscle pain. Ciguatera causes numbness, burning, or tingling of the hands and legs, or around the mouth. In two to five days,
victims often experience an intense, painful tingling or burning sensation when touching cold objects. Because this classic symptom is often
delayed, it should not be relied on for making an early diagnosis. In
Hawai'i, itching and hives are common; in some parts of the Pacific,
these symptoms are rare.

Other less common symptoms include chills, dizziness, sweating,
headache, and a metallic taste in the mouth. A slow heart rate with low

Bluefin trevally, or *'ōmilu (Caranx melampygus),* are members of the Jack family. Jacks often feed on reef fish, making them frequent carriers of ciguatera
poisoning. Hanauma Bay. (D. R. and T. L. Schrichte)

blood pressure, or a fast heart rate with high blood pressure, may appear, along with decreased reflexes and dilated pupils.

Nausea, vomiting, abdominal cramps, and diarrhea last for one to two days. Weakness may last for seven days. Neurologic symptoms can reoccur for a month or more. Some researchers believe ciguatoxin and other related marine toxins accumulate in humans, as in fish. Because of this build-up, habitual fish eaters may show symptoms after eating only a small amount of toxic fish.

Even though some people seem to be less sensitive to the toxin than others, no one is immune.

Usually, no single patient develops every symptom, but most experience a unique tingling sensation in response to cold.

Months after a poisoning, long-term symptoms of muscle aches, joint pain, and a weak, tired feeling can appear when a victim eats fatty foods, seafood, or drinks alcohol.

First Aid

Most cases of ciguatera are neither severe nor life threatening. Some symptoms may be relieved by intravenous mannitol, available in emergency rooms.

For persistent hives and itching, try taking one or two 25-milligram diphenhydramine (Benadryl) tablets every six hours. This drug is sold without a prescription. Diphenhydramine may cause drowsiness: Do not drive, swim, or surf after taking this medication. For muscle aches and headaches, try two 200-milligram ibuprofen tablets with food every six hours.

For severe ciguatera illness, see a doctor immediately for removal of remaining toxin from the digestive tract and treatment of symptoms. Paralysis or difficulty breathing is always a medical emergency. Take the remains of the suspect fish with you (including head and guts) or freeze it for later examination. Stop anyone else from eating the fish. Report any ciguatera incidents, even minor, to the Hawaii Department of Health to help prevent further illnesses.

Advanced Medical Treatment

No specific antidote exists for ciguatera poisoning. Intravenous mannitol (20 percent), 1 gram per kilogram given over about thirty minutes, and within six days of ingestion, will reduce the number of symptoms both in mild and severe poisonings. Patients treated with mannitol have fewer symptoms within two and a half hours after treatment.[1] Correct dehydration before giving mannitol. Mannitol infusion may be helpful even weeks after the initial poisoning.

No clinically important tests exist for acute diagnosis. This poisoning is not usually life threatening. In severe cases, however, respiratory paralysis and bronchospasm can cause death.

Treatment is supportive. For severe cases, initiate life support.

Patients rarely arrive at the emergency room early enough for lavage to be useful. If the ingestion has been within one hour, empty the stomach. After lavage, or in patients greater than one hour after eating the fish, give oral charcoal. For hypotension, give IV fluids and dopamine. For bradycardia, give atropine.

Hydroxyzine may help itching, which can persist for weeks. Steroids do not appear to be beneficial.

Since this toxin accumulates in tissues, instruct patients to avoid repeated exposure. Warn them that symptoms may recur after eating seafood, fatty foods, or drinking alcohol.

Report all cases to the Hawaii Department of Health. Send a specimen of the suspect fish for testing.

FISH HANDLER'S DISEASE (*ERYSIPELOTHRIX RHUSOPATHIAE*)

Fish handler's disease is an infection caused by *Erysipelothrix rhusopathiae,* a bacterium commonly found in seawater and on the skin of fish and invertebrates. This robust organism, which also lives in fresh water, tolerates drying, salting, pickling, or smoking. Even after days of exposure to direct sunlight or months in rotting flesh, the bacterium survives.

Mechanism of Injury

Erysipelothrix rhusopathiae usually enters a person's skin through an animal bite, a fish-fin puncture, or a wound made by a seashell. The infection occasionally enters the bloodstream and spreads throughout the body.

Incidence

Fish handler's disease is not common in Hawai'i, but it has been reported. *Erysipelothrix rhusopathiae* recently infected a woman here after she had been cleaning fish. She developed bacterial endocarditis, a serious infection of the heart.

Fish handler's disease is an infection caused by a specific bacterium *(Erysipelothrix rhusopathiae)* that gets into a cut. Because these bacteria are common on the skin of fish, anyone can get the infection. This fish auction worker in Hilo is at higher risk than most people. (Susan Scott)

People who handle fresh fish or crabs are at risk for this infection, which is common in some coastal areas of the United States.

Prevention

Do not touch fish or crabs if you have an open hand wound. If you are bitten, cut, or punctured by any kind of marine animal, wash the wound thoroughly. Meticulous scrubbing helps reduce the risk of marine infections.

Signs and Symptoms

This infection has a distinct appearance. Two to seven days after infection, a purplish, swollen area appears around the puncture wound. Around this purplish area, a clear area appears in a circle, surrounded by another raised, purplish, swollen area.

The area is usually warm to the touch and hurts and itches. Often, the puncture wound drains. Lymph nodes of the affected arm may swell. The infection may spread down the finger, into the web space, and up another finger. The palm is not usually affected, but the infection may spread to the wrist or forearm. Fever and a feeling of overall illness may develop. The original puncture wound may enlarge, and other lesions can appear in other areas of the hand.

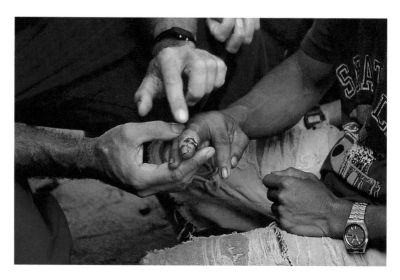

This fisherman in Zambales, the Philippines, punctured his finger with a fish fin while landing a catch. A week later, his finger had the distinctive appearance of *Erysipelothrix rhusopathiae*. (Vernon Ansdell, M.D.)

 **First Aid**

Fish handler's disease needs antibiotic treatment as outlined below in the section on Advanced Medical Treatment.

An advanced infection, especially with fever and chills, can be a medical emergency. Seek help immediately. (See *Staph, Strep,* and General Wound Care.)

 **Advanced Medical Treatment**

This usually localized infection is sensitive to penicillin. Alternatives are cephalexin or ciprofloxacin. Continue antibiotics until the infection resolves. Without treatment, most cases clear up in about three weeks. The above antibiotics, however, treat the infection rapidly.

Rarely, the infection spreads to the joints, producing septic arthritis, or becomes blood-borne, causing endocarditis. These patients need intravenous antibiotics, such as penicillin G, for up to six weeks.

For serious infections, culture the wounds. Alert the lab that you suspect *Erysipelothrix rhusiopathiae*. This gram-positive bacillus can be confused with *Streptococci* or *Diphtheroids*.

Clean and debride the wound, and check for foreign bodies with soft-tissue x-ray or ultrasound. Immunize for tetanus as detailed in *Staph, Strep,* and General Wound Care.

FISHWORM INFECTIONS (ANISAKIASIS)

Parasites infect nearly every kind of marine animal. Although most of the parasites do not infect humans, one type of parasitic worm infection is currently on the rise in the United States. Called anisakiasis, this infection comes from eating raw fish or squid.

In Hawai'i, marine worm infections come mainly from *Anisakis simplex,* a worm with a life cycle beginning and ending in the digestive tract of whales, dolphins, and seals.

The cycle starts when a female worm lays eggs in the host's stomach. The eggs pass through the host's digestive tract, into the feces, and eventually end up in the water. The eggs hatch into juveniles, which are eaten by drifting, shrimplike crustaceans. Fish and squid eat the infected shrimp. When a whale, dolphin, or seal eats an infected fish or squid, the worm's cycle is complete. When a human eats an infected animal, however, the cycle is broken. Juvenile worms cannot mature in humans, but the worms can cause illness before they die.

Mechanism of Injury

Juvenile *Anisakis simplex* worms, about one twenty-fifth of an inch thick and up to 1 inch long, often cause few or no symptoms before they die and pass harmlessly through the human digestive tract. Occasionally, the worms bore into, or through, the stomach or intestinal wall, causing severe pain, intestinal perforation, and even death.

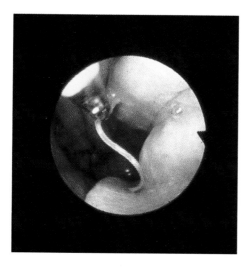

A forceps is being used to pull an *Anisakis simplex* worm (a fish parasite) from a human stomach. When eaten alive, these burrowing worms can cause severe pain. (T. Fukumura, M.D.)

Raw salmon, as in this *lomilomi* salmon dish, has been the major source of Hawai'i's *Anisakis* fishworm cases. (Susan Scott)

Incidence

Eight cases of anisakiasis have been reported in Hawai'i. More cases probably have occurred but either presented no symptoms or were not correctly diagnosed.

Imported raw salmon (used in sushi and lomilomi salmon) was the major source of Hawai'i's cases; raw squid and either yellowfin tuna (*ahi,* or *Thunnus albacares*) or skipjack tuna (*aku,* or *Katsuwonus pelamis*) have also caused this parasitic infection.

Six of Hawai'i's cases developed after home-prepared meals; two came from a sushi bar. In Japan, about one thousand cases of anisakiasis are reported each year. At least two deaths have been reported worldwide.

Prevention

Freezing or cooking kills these worms, thus preventing infection. *Anisakis* worms are difficult to see in the animals' flesh. Careful inspection or "candling" of a fillet does not reveal all the worms.

The risk of people in Hawai'i becoming infected from raw, locally caught fish is extremely low. Raw, imported fish or squid caused six of Hawai'i's eight diagnosed cases.

Signs and Symptoms

Some people experience little or no stomach pain after eating these worms. In some cases, though, severe, intermittent stomach pain, sometimes with nausea and vomiting, begins from one hour to fourteen days after eating the infected fish or squid. Some victims also develop diarrhea. Persistent, severe pain may be a sign of intestinal perforation or widespread infection.

Hives and one or more swollen joints may be signs of an allergic reaction to a worm embedded in the lining of the digestive tract.

First Aid

For any severe stomach pain, see a doctor immediately.

Advanced Medical Treatment

No antihelminthic drug exists for this parasite. A specific diagnostic assay exists but is not readily available. Stool tests are not useful, since this parasite does not mature and release eggs in its human host.

Definitive diagnosis, and treatment, is by endoscopy.[1] Treatment is removing the worms with endoscopic biopsy forceps. If the worms can't be retrieved, they usually die spontaneously. If bowel perforation or obstruction occurs, surgery is necessary. This illness can be life threatening. Peritonitis, following perforation of the bowel at the burrowing site, has caused at least two deaths worldwide.

Suspect anisakiasis in anyone who has eaten raw fish or squid. This clinical presentation is often confusing. Cases of suspected appendicitis, diverticulitis, cholecystitis, tuberculous peritonitis, gastrointestinal tumors, and Crohn's disease have turned out to be anisakiasis.

Eosinophilia (up to 41 percent) occurs in chronic stages of this illness.[2] These patients often have hives, itching, and swollen joints. Those with joint swelling are often diagnosed with inflammatory arthritis. The symptoms do not respond to anti-inflammatory drugs but disappear when the worm is removed from the digestive tract.[3]

Ultrasound shows thickening of the small intestinal wall. Upper GI series may show threadlike filling defects. Report cases of fishworm infection to the Hawaii Department of Health.

HALLUCINATORY FISH POISONING

In Hawai'i, two species of mullet and two species of goatfish have been implicated in a poisoning that causes temporary illness with hallucinations.

Certain surgeonfish (family Acanthuridae) and chubs (family Kyphosidae) may also cause this illness, but reports about them are inexact.

Mechanism of Injury

This heat-stable toxin is concentrated in the heads of the fish. Its action is unknown. Only certain fish contain this toxin, presumably from

Sources of Hallucinatory Fish Poisoning in Hawai'i

English Name	Hawaiian Name	Scientific Name
Striped mullet	*'ama'ama*	*Mugil cephalus*
Acute-jawed mullet	*uouoa*	*Neomyxus leuciscus*
Yellowstripe goatfish	*weke'ā*	*Mulloides flavolineatus* (formerly *M. samoensis*)
Band-tailed goatfish	*weke pahulu*	*Upeneus taeniopterus* (formerly *U. arge*)

something they eat. When brains of suspect fish were fed to a cat, it "at once went crazy."[1]

Incidence

This relatively rare poisoning is most common during the summer months, coming from fish caught near Moloka'i, Kaua'i, and O'ahu. No fatalities have been reported. No cases of hallucinogenic fish poisoning were reported in 1991, 1992, or 1993. Three cases were reported in 1994, one case in 1995.

During the Cold War, incidence and location data about this syndrome were suppressed, because "Russia was exceedingly interested in nerve drugs such as this."[2]

Eating the heads of yellowstripe goatfish, or *weke'ā* (*Mulloidichthys flavolineatus*), sometimes causes temporary hallucinations. Kahe Point. (D. R. and T. L. Schrichte)

Eating the heads of some mullets has occasionally caused hallucinatory fish poisoning. Affected fish look normal. Hanauma Bay. (D. R. and T. L. Schrichte)

Prevention

Do not eat the heads of implicated mullet and goatfish caught near Moloka'i, Kaua'i, or O'ahu in June, July, or August.

Signs and Symptoms

Symptoms develop within five to ninety minutes. They include tingling around the mouth, sweating, weakness, hallucinations, and chest tightness. The toxin affects some people during sleep, producing vivid nightmares. Because of this, some people call the band-tailed goatfish "the nightmare *weke*." Hawaiians called this same fish *weke pahulu* (chief of the ghosts).

 ## First Aid

Victims hallucinating or extremely depressed should go to an emergency room for help. Victims not hallucinating or extremely depressed should remain calm and wait until the symptoms disappear, usually overnight.

Save any remaining fish for analysis. Report all cases to the Hawaii Department of Health.

 ## Advanced Medical Treatment

No antidote or specific therapy exists for this poisoning. It is not life threatening. Evaluate the patient for other causes of psychosis. Hallucinating patients may need sedation. For recent ingestion, empty the stomach and give charcoal. Save any specimens of the fish for testing by the Hawaii Department of Health.

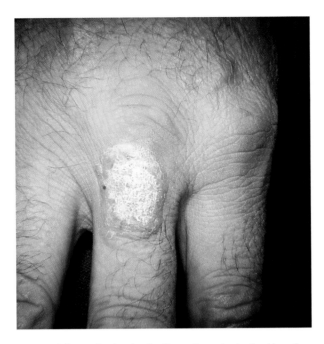

This nodule on the back of a finger is typical of a *Mycobacterium marinum* infection. Three or four weeks can elapse after a cut or scrape before the nodule appears. (Allan Izumi, M.D.)

MYCOBACTERIUM MARINUM

Mycobacterium marinum, formerly called *Mycobacterium balnei,* was discovered in the 1950s as the organism causing elbow infections in swimmers using a swimming pool in Sweden. Because of this, an infection from *Mycobacterium marinum* was called a "swimming pool" granuloma.

Since then, *Mycobacterium marinum* has been found commonly in aquariums, lakes, streams, oceans, and swimming pools. Today, infections from this organism are sometimes called "fish-tank" granulomas, although most often the bacterium is known by its scientific name, *Mycobacterium marinum. Mycobacterium* is the name of a group of bacteria well known in medicine. Other members of this group cause tuberculosis (TB) and Hansen's disease (leprosy).

Mechanism of Injury

This localized infection is caused by the organism entering the body through a cut or scrape.

Incidence

In one report, eight cases of *Mycobacterium marinum* in Hawai'i were all originally misdiagnosed. The source of infection in some of these cases was irrigation ditches in sugarcane fields. This infection recently developed in a Hawai'i aquarium worker after a sea urchin *(wana)* punctured her finger while she was cleaning a fish tank. This case was also originally misdiagnosed.

Nationally, from 1963 through 1985, 590 cases of *Mycobacterium marinum* were reported. In Queensland, Australia, twenty-nine people were infected from 1971 to 1990.

At highest risk are agricultural workers, fishermen, swimmers, and aquarium workers. One case involved a British aquarium worker after a dolphin bite. *Mycobacterium marinum* is not usually passed from person to person.

Prevention

If you have a cut on your finger or hand, avoid cleaning fish tanks or immersing your hands in obviously dirty water. Also, people with damaged immune systems, such as those with diabetes, liver disease, splenectomy, AIDS, valvular heart disease, and anyone taking steroids, should be aware that they are at higher risk for contracting infections.

After any marine wound, meticulous scrubbing helps reduce infection risk. (See *Staph, Strep,* and General Wound Care.)

Chlorine in swimming pools does not destroy *Mycobacterium marinum.*

Signs and Symptoms

Symptoms of *Mycobacterium marinum* can begin seven to ten days after a puncture wound or scrape. Three to four weeks can elapse, though, before a nodule appears. This infection sometimes heals on its own in two to three years but often leaves scarring and localized skin discoloration.

Eighty to 90 percent of *Mycobacterium marinum* cases are on a person's finger, hand, arm, or elbow. Because this organism grows best in temperatures below about 90 degrees F, most infections are on the fingers and toes.

The infection can have several appearances. One is a raised, red bump, growing to 1/2 to 1 inch wide, near the site of infection. This lesion can be hard and scaly, sometimes appearing purple at the base. Occasionally, the bump will open and drain. It can grow to be more than 2 inches wide. *Mycobacterium marinum* can also appear as a line of bumps, or abscesses, running up the affected limb along superficial lymph vessels.

A third, and rare, sign of this infection is multiple skin lesions on the cooler parts of the body, such as the face, arms, and legs. This is a blood-borne *Mycobacterium marinum* infection, sometimes seen in victims of AIDS or in people with other immune deficiencies.

First Aid

No first aid treatment exists for these lesions. If you are in a high-risk group for *Mycobacterium marinum* and have any of the above symptoms, see a doctor. Describe, if you can, how you may have been exposed to this organism.

Advanced Medical Treatment

Patients with *Mycobacterium marinum* typically have had a chronic, nonhealing sore for several months, with a history of work or recreation in water. Misdiagnosis is common. In a California study, the mean delay to diagnosis was ten weeks.[1]

Confirm the diagnosis with a punch skin biopsy. Use standard *Mycobacterium* medium at 31 to 33 degrees C. Acid-fast bacteria are visible in approximately 10 percent of biopsies. About 50 percent of cultures are positive. Radiometric assays improve sensitivity. It is

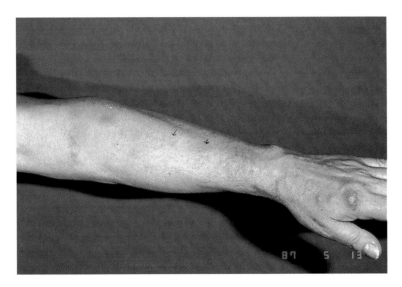

Occasionally, *Mycobacterium marinum* appears as a line of bumps running up the limb, like the ones in this photo. The primary lesion is on the hand. (Steve Ostroff, M.D., Centers for Disease Control)

important to consider *Mycobacterium marinum* in patients with mildly reactive tuberculin skin tests.

Antibiotic treatment for *Mycobacterium marinum* has not been well studied. Long duration of therapy appears to be necessary. Treat with rifampin, 600 milligrams per day, and ethambutol, 1,200 milligrams per day for six weeks or until all lesions disappear. Dosages and duration vary widely in the literature.[2] Doxycycline, trimethoprim-sulfa, clarithromycin, and ciprofloxacin may also work. Most lesions respond to antibiotics. Do not use steroids.

Surgery may be necessary when deep structures of the hand are involved, impairing function. Immune-compromised patients can have widespread, rapidly progressive infection.

PUFFERFISH POISONING (TETRODOTOXIN)

Tetrodotoxin is a potent poison common in species of several fish families: pufferfish (Tetraodontidae), porcupinefish (Diodontidae), and ocean sunfish (Molidae). Tetrodotoxin is distributed throughout fishes' bodies differently in different species. Potency of the toxin varies through the seasons.

This toxin is also found in other marine animals, such as certain seastars, crabs, snails, shells, worms, Australia's blue-ringed octopus, and even some species of red calcareous algae. The skin of some frogs and newts contains tetrodotoxin.

Most animals use their tetrodotoxin for protection only. For example, the pufferfish common in Hawai'i often secrete considerable amounts of tetrodotoxin into the surrounding water when disturbed. The blue-ringed octopus, and some species of ribbon worms, are exceptions. These animals use their tetrodotoxin to immobilize prey.

Because tetrodotoxin is found in such a wide range of animals, some researchers believe bacteria living on, or in, their host produce the poison. This hypothesis is supported by the fact that pufferfish reared in captivity have no tetrodotoxin in their tissues. Also, when newts' skin is cleared of toxin and bacteria in the laboratory, tetrodotoxin does not reappear. Some common bacteria, such as *Escherichia coli*, *Pseudomonas* species, and *Vibrio* species, are capable of producing tetrodotoxin.

In Japan, pufferfish is a delicacy. The favored types to eat there are of the genus *Fugu* (not found in Hawai'i), thus the name *fugu* for this potentially lethal dish.

Like its pufferfish relatives, this porcupinefish, or *kōkala (Diodon hystrix),* carries tetrodotoxin in its flesh and organs. Hanauma Bay. (D. R. and T. L. Schrichte)

People eat fugu to experience a sense of euphoria. Whether this comes from eating minute amounts of tetrodotoxin or from the exhilaration of tempting death remains in question.

Mechanism of Injury

Tetrodotoxin poisoning is from eating affected fish. No cases have been reported of tetrodotoxin poisoning from seawater.

Tetrodotoxin is a potent nerve poison, causing paralysis by blocking nerve conduction. This nonprotein toxin interferes with the movement of sodium into nerve and muscle cells. Tetrodotoxin is 160,000 times as potent as cocaine at blocking nerve conduction.

Tetrodotoxin acts directly on the brain stem, causing vomiting, and may directly depress the respiratory center. It also directly blocks action on skeletal muscle fibers and interferes with blood clotting.

Incidence

Hawai‘i's most common pufferfish, the stripebelly pufferfish *(Arothron hispidus)* also known as *kēkē* or *‘o‘opu hue,* has caused at least seven deaths in Hawai‘i.

It is illegal for anyone to serve pufferfish in Hawai‘i. Rumor has it, though, that some restaurants here secretly serve fugu to people who want it. Also, some Hawai‘i residents bring home packaged, dried fugu from Japan. Three California residents recently were poisoned by imported Japanese fugu.

Earlier in this century, more than one hundred fugu deaths a year were reported in Japan. That number has dropped to about forty-five cases, resulting in about five deaths a year over the past decade. This dramatic decrease in the tetrodotoxin death rate is probably because chefs in Japan must now pass rigorous licensing tests to prepare fugu. Since this regulation, almost no cases of poisoning have resulted from eating fugu in Japan's restaurants. The poisonings that have been reported there involved inexperienced fishermen who prepared their own fish.

Prevention

Do not eat any kind of pufferfish, or any other animal reported to contain tetrodotoxin. Tetrodotoxin is neither destroyed nor inactivated by cooking, canning, drying, freezing, or smoking.

Signs and Symptoms

Most people who eat fish containing tetrodotoxin either have no poisoning or only mild poisoning. Symptoms usually begin ten to forty-five minutes after eating but may be delayed up to three hours. In general, the faster the onset of symptoms, the more severe the poisoning. Prognosis is good if the patient survives the first twenty-four hours.

Serious pufferfish poisoning progresses as follows:

1. Within five to forty-five minutes, the person feels numbness or unusual sensations around the mouth. This is the classic symptom of tetrodotoxin poisoning. Nausea with no vomiting is common.

2. In ten to sixty minutes, the numbness moves to the tongue, face, and other areas of the skin. The person has slurred speech and poor coordination.

3. In fifteen minutes to several hours, a person with severe poisoning has widespread paralysis with low blood pressure, and difficulty talking and breathing. Even though the pupils may be fixed and dilated, and the person is unable to respond, most people at this stage are still conscious. They can remain so throughout the entire episode.

4. The final stages of pufferfish poisoning can occur from fifteen minutes to twenty-four hours. The breathing muscles become paralyzed, and the heart beats slowly and irregularly. The person may or may not be unconscious. Death is usually from breathing failure. One person died only seventeen minutes after eating pufferfish.

 **First Aid**

Pufferfish poisoning is a medical emergency. Take the person with symptoms to the nearest emergency room immediately. Never induce vomiting. Even partial paralysis of the throat could cause the victim to inhale vomited material, which can result in death. Begin cardiopulmonary resuscitation if breathing stops.

Stop anyone else from eating suspect fish. Because of the deadliness of this toxin, others who may have eaten any of the suspect fish should also go to an emergency room. Take with you, or freeze, the remains of the fish for testing.

Advanced Medical Treatment

No antidote exists for tetrodotoxin, and no clinically useful tests exist for diagnosis. The mortality rate for tetrodotoxin poisonings is approximately 11 percent. The primary cause of death is respiratory collapse.

Treatment is supportive. Since stomach emptying may be delayed, lavage any victim who arrives within three hours of eating the fish. The toxin may be partially inactivated by sodium bicarbonate.[1] Lavage with dilute sodium bicarbonate. After lavage, give charcoal. Seriously affected patients often need intubation.

Complications include paralysis, vomiting with aspiration, hyper- and hypotension, arrhythmias, coagulopathies, and seizures. No proven drug therapy exists. Use standard therapies for arrhythmias, blood pressure, and seizure control.

Report all cases to the Hawaii Department of Health. Send fish specimens for testing.

SARDINE POISONING (CLUPEOTOXIN)

Fish in the order Clupeoformes are herrings, sardines, anchovies, tarpons, and bonefishes. Hawai'i hosts several native fish belonging to this order. A foreign species, the Marquesan sardine *(Sardinella marquesensis),* was introduced to Hawaiian waters in 1955 as a baitfish for tuna fishermen. Rarely, some members of the order Clupeoformes become extremely toxic.

Marquesan sardines *(Sardinella marquesensis),* imported to Hawaiian waters in 1955 as baitfish, caused the death of a Kaua'i man in 1978. Three of these fish were mixed with a catch of *akule,* or big-eyed scad *(Selar crumenophthalmus).* The man ate all three of the sardines and died fifteen hours later, in spite of early hospitalization. (John E. Randall)

Mechanism of Injury

Because of the rarity of this acute illness, researchers have identified neither the toxin nor its action. Apparently, these fish eat toxic organisms, perhaps dinoflagellates, that do not visibly injure the fish. The toxin is found in the fish's guts.

Clostridium perfringens, a bacterium marker for human fecal bacterial contamination in water, has been cultured from *Sardinella marquesensis,* a Marquesan sardine associated with clupeoid poisoning. The significance of this is unclear.

Incidence

In 1978, the sardine *Sardinella marquesensis* caused the death of a sixty-seven-year-old Kaua'i man who had been in excellent health. He became ill thirty minutes after eating the fish (steamed for three hours) and died fifteen hours later despite being admitted to the hospital early in the course of the illness.

Clupeoid poisoning occurs rarely and sporadically in the South Pacific, Indian Ocean, and Caribbean. Worldwide incidence, although always low, is higher during summer. Most poisonings are from fish caught close to shore. No clupeotoxin illnesses have been reported from canned fish.

Prevention

Avoid eating fresh fish from this order, especially the guts, during summer months. Toxic fish appear normal. This toxin does not create any unusual flavor, smell, or appearance. Cooking did not deactivate this toxin in the Kauaʻi case.

Signs and Symptoms

Symptoms of this poisoning may begin within fifteen minutes after eating the fish. Victims often experience a metallic taste and dryness of the mouth immediately after eating the fish. This progresses to nausea, vomiting, diarrhea, abdominal pain, and collapse. Nervous system symptoms are dilated pupils, violent headaches, numbness, tingling, excess salivation, muscle cramps, breathing difficulty, paralysis, convulsions, and coma.

The mortality rate is up to 45 percent. Death can take place within fifteen minutes. Some deaths have been so rapid that remnants of the fish were still in the victim's mouth.

 ## First Aid

Because of the severity of this illness, rush any suspected victim of clupeotoxin to the nearest emergency room.

 ## Advanced Medical Treatment

Clupeotoxin is life threatening. No antidote or specific therapy exists. Empty the stomach and administer charcoal. Victims may have multisystem organ failure, requiring ventilatory and circulatory support.

Report suspected cases to the Hawaii Department of Heath.

SCOMBROID POISONING

Scombroid is a poisoning that usually comes from eating gamefish not properly refrigerated before cooking or preserving. When the fish become warm, bacteria on the skin begin to break down the fishes' flesh, normally laden with the amino acid histidine. During the breakdown process, histidine turns into histamine and saurine, a salt of histamine. Histamine is a major chemical messenger for allergic reactions.

Normally, eating histamine causes no ill effects in humans. It is poorly absorbed by the stomach and, in the liver and bowel, converts to

Scombroid poisoning can result when freshly killed fish are not kept cold. A fisherman delivered these yellowfin tuna *(ahi)* to the Hilo fish auction, where they were immediately placed on ice. (Susan Scott)

another, nontoxic form. Some researchers believe histamine is not the only culprit in scombroid poisoning. Other chemicals in minimally spoiled fish, such as cadaverine or putrescine, may help histamine pass into the bloodstream by blocking enzymes that normally break down histamine.

The word *scombroid* comes from the family name of tunas and wahoo *(ono),* Scombridae. Other fish family members, however, cause this poisoning at least as often as tunas. Because of this, one researcher suggests calling this poisoning "pseudoallergic fish poisoning," a more appropriate name.

From 1989 through 1993, more than half the reported cases of scombroid in Hawai'i came from dolphinfish, or *mahimahi (Coryphaena hippurus),* and yellowfin tuna, or *ahi (Thunnus albacares).* The rest came from billfish, or *a'u* (Istiophoridae); skipjack tuna, or *aku (Katsuwonus pelamis);* jacks, or *ulua* (Carangidae); moonfish, or *opah (Lampris regius);* mackerel scad, or *'ōpelu (Decapterus pinnulatus);* big-eyed scad, or *akule (Selar crumenophthalmus);* snappers; and salmon.

Not only fish can cause scombroid poisoning. Rarely, aged cheddar, Swiss, and gouda cheeses have caused this poisoning.

Mechanism of Injury

Whether histamine is released by the body in reaction to a substance or eaten in spoiled fish, the results are the same: blood vessels dilate and leak fluid into the surrounding tissue. On the skin, this produces a red, raised rash; in the respiratory tract, wheezing; and in the stomach, vomiting.

From 1989 through 1993, more than half the reported scombrid poisoning cases in Hawai'i came from *mahimahi,* or dolphinfish *(Coryphaena hippurus).* United Fishing Agency Auction, Honolulu. (D. R. and T. L. Schrichte)

The normal level of histamine in fish is less than 0.1 milligram per 100 grams of flesh. Histamine levels between 10 and 20 milligrams per 100 grams of flesh may make people ill. Toxic fish can have histamine levels of up to 600 milligrams per 100 grams of flesh.

Incidence

In Hawai'i, an average of about fifty cases of scombroid are reported a year. Seventy-eight cases were reported in 1995. New York, California, and Hawai'i have the highest number of cases in the country. In the United States, scombroid poisoning accounts for 5 percent of all food-borne outbreaks.

True allergies to fish and other seafood do occur, but since scombroid poisoning mimics allergy, some people falsely believe they are allergic to fish.

Prevention

To prevent scombroid poisoning, ice all freshly caught fish immediately. Never leave fish carcasses lying on a deck or beach in the hot sun.

Scombroid poisoning can be avoided if fish are stored at 59 degrees F (15 degrees C) or lower.

Some affected fish have a peppery or sharp metallic taste, but others may look, smell, and taste normal. Not everyone eating a scombroid toxic fish becomes ill, probably because the decay in the fish is uneven. If one person appears to be allergic to a fish, others should stop eating it.

The toxins causing scombroid poisoning are neither destroyed nor inactivated by cooking, canning, drying, freezing, or smoking.

Signs and Symptoms

Symptoms of scombroid poisoning begin five to ninety minutes after eating the affected fish and resemble a moderate to severe allergic reaction: flushing (particularly of the face), a sensation of warmth, red eyes, itching, hives, swelling of the face and lips, wheezing, nausea, vomiting, diarrhea, stomach pain, abdominal cramping, difficulty speaking, thirst, sore throat, burning of the gums, increased heart rate, dizziness, and low blood pressure. A throbbing headache is common.

Without treatment, symptoms usually disappear in eight to twelve hours. Shock and death from severe cases of scombroid poisoning have been documented in the past but have not been reported in recent years. A hoarse voice, a feeling of thickening of the throat, or difficulty breathing can be life-threatening signs either of allergy or scombroid poisoning.

People taking isoniazid (INH), a tuberculosis medicine, may have more severe symptoms. INH blocks natural enzymes in the body that break down histamine. Monoamine oxidase inhibitors, a type of antidepressant, also may increase the severity of scombroid poisoning.

First Aid

For mild to moderate scombroid reactions, take diphenhydramine (Benadryl), one to two 25-milligram tablets every six hours. This over-the-counter drug causes drowsiness: Do not surf, swim, or drive after taking it.

Scombroid poisoning can cause breathing difficulties. Because of this, consider any unusual breathing a medical emergency and go directly to an emergency room. Take or freeze any remaining fish for laboratory examination.

Epinephrine (adrenaline) is the emergency treatment for breathing distress due to either scombroid poisoning or allergy. This injectable medication comes in an auto-injector, available by prescription. Dive leaders and anyone who has had a previous, severe allergic reaction to anything should carry one on all ocean outings. Administer the epinephrine as soon as a scombroid or allergy victim begins wheezing or complains of tightness in the chest.

Epinephrine is a potent medication that causes rapid heartbeat, anxiety, and high blood pressure. Rarely, it can cause heart attacks or strokes. Reserve this medication for victims with breathing problems.

Advanced Medical Treatment

No clinically useful tests exist for the diagnosis of scombroid poisoning. Life-threatening reactions are uncommon. No deaths from scombroid poisoning have been reported in Hawai'i. Differentiating scombroid poisoning from true allergy is not important for initial treatment.

Antihistamines are effective initial treatment. Hydroxyzine, 50 milligrams intramuscularly or orally, is slightly more effective than diphenhydramine. For a persistent rash or headache resistant to antihistamines, use cimetidine.[1]

For patients with bronchospasm, inhaled therapy with a beta-agonist, such as albuterol, is often effective. For persistent wheezing or rash, give 0.3 milligrams epinephrine subcutaneously. Most patients improve rapidly, being symptom free in three to thirty-six hours.

For severely ill patients who have eaten the fish within one hour, use gastric lavage followed by charcoal. Give other patients charcoal only. Rarely, poisoning victims develop systemic shock requiring intravenous saline for hypotension. For severe hypotension or bronchospasm not responding to initial therapy, try epinephrine, 0.1 milligram intravenously. The role of steroid therapy is uncertain.

Send suspect fish to the Hawaii Department of Health for histamine testing. The Department of Health currently judges bonito, *aku,* and albacore positive at 10 milligrams per 100 grams of fish and *mahimahi* positive at 25 milligrams per 100 grams of fish.

If histamine levels in a suspect fish are high, be sure to inform victims that they are not allergic to fish. Explain how to avoid scombroid poisoning in the future.

SEAL FINGER

Hawai'i's only native seal is the Hawaiian monk seal *(Monachus schauinslandi),* an animal named for its solitary habits and for its wrinkled neck, which resembles a monk's cowl. Female monk seals are

Seal finger is a rare infection caused by a bite or scratch from a seal. In the past, seal finger mainly afflicted people who were killing seals. Today, the infection is more likely in people trying to save seals, such as this worker feeding a Hawaiian monk seal at Sea Life Park, O'ahu. (Monte Costa)

slightly larger than males, growing to more than 7 feet long and weighing nearly 500 pounds. These animals spend most of their time at sea, using their sharp teeth to catch fish, octopus, and lobsters. Monk seals often rest, and always give birth, on beaches.

No one knows the abundance of monk seals in Hawai'i before humans arrived, but the animals are now scarce. A federally protected species, it is in critical danger of extinction.

Beachgoers occasionally encounter a monk seal on a main island beach. Approaching within 100 feet of any monk seal is a federal crime. Not only do these solitary creatures need to rest and nurse their young on the beach, they may bite if startled.

Seal finger is a rare infection resulting from a bite or a scratch from a seal. A Norwegian doctor first described this condition, at the time called *spaick* finger, or blubber finger, in a medical journal in 1907. (*Spaick* is the Norwegian word for blubber.) At that time, and later in the century, men who killed and skinned seals occasionally suffered this condition on a finger or hand.

Today, most countries protect seals, but that has not stopped the incidence of seal finger. The disease now occurs mainly on the hands of people trying to save seals rather than kill them.

Mechanism of Injury

Researchers believe the organism causing seal finger may be a mycoplasma bacteria, but its exact identity is unknown. The infectious agent enters the finger through a minor cut made by the teeth or claws of the seal. Seal finger can also result from touching a seal's fur with a finger or hand bearing a previous cut.

Incidence

This disease is most frequent in bites to workers in marine parks and wildlife rehabilitation facilities. Although seal bites are relatively common among monk seal workers in Hawai'i, no cases of seal finger have been reported here.

Prevention

Workers with cuts on their fingers or hands should avoid handling seals until the cuts are thoroughly healed. When possible, wear gloves while feeding, tagging, or otherwise caring for seals.

If you find a monk seal basking on a beach, go back the way you came or pass widely around the animal. Do not disturb it in any way. It is best not even to let the seal see you. Notify the National Marine Fisheries Service (or any other federal or state wildlife officials) of the sighting.

You can help the seal, and protect other seals, by asking people to stay well away from the animal. Never throw water on a monk seal or chase it back into the water. Strictly supervise young children who may rush toward the seal.

Signs and Symptoms

Symptoms of seal finger begin from three to twenty-one days after contact. The affected finger becomes extremely painful, swollen, and tender. Lymph nodes in the affected arm may become swollen, but otherwise, the infection does not spread through the bloodstream or cause overall illness.

Without prompt, proper therapy, adjacent fingers may also become infected. Permanent, debilitating joint injury is possible with delayed treatment.

First Aid

After any seal bite or scratch, wash the wound vigorously. Gently pull the edges of the skin open and scrub directly inside the wound with gauze or a clean cloth soaked in clean, fresh water. For bites that bleed, press a clean cloth directly against the wound until the bleeding stops. If bleeding persists or the edges of a wound are jagged or gaping, the victim probably needs stitches. Taping a cut shut is often effective but may leave a more visible scar than suturing. For more details on wound care, see *Staph, Strep,* and General Wound Care.

If a finger or toe is numb or will not move normally after a bite, or for any sign of infection, see a doctor. Mention your contact with seals, a fact important for proper therapy.

Advanced Medical Treatment

Thoroughly scrub, explore, irrigate, and debride all wounds. Examine for tendon and nerve damage and repair, or refer, as necessary. Suturing wounds to control bleeding, preserve function, or improve appearance is appropriate.

For seal finger, use tetracycline, 500 milligrams four times a day until the infection is gone.[1] Doxycycline, 100 milligrams twice a day, is a good alternative. Delay in this treatment can cause permanent joint damage, requiring arthrodesis. Cephalosporins and penicillins are not effective.

Seals can transmit several other infections, such as fish handler's disease (see Fish Handler's Disease) and leptospirosis. Leptospirosis spirochetes, found in Hawai'i's streams and estuaries, enter humans via a break in the skin or across intact mucous membranes. Doxycycline, tetracycline, or penicillin are treatments for leptospirosis.

SEAWEED POISONING

Seaweed, called limu, is a popular food item in Hawai'i. One kind, *limu manauea,* sometimes called short *ogo (Gracilaria coronopifolia),* is a red alga growing on reef flats, and occasionally in tide pools. These red-to-pinkish plants grow to about 6 inches tall. Fifty years ago, *limu manauea* was common along the shores of O'ahu. People ate this limu so much that it became overpicked and is now scarce.

People prepare *limu manauea,* or short *ogo,* several ways in Hawai'i. One is the Japanese-style *ogo namasu,* combining blanched seaweed with vinegar, sugar, soy sauce, and sliced cucumbers. Another common dish is spicy *ogo kim chee,* made with garlic, onions, and chili peppers.

Some people have become ill in Hawai'i after eating *limu manauea* harvested from the wild. Commercially cultivated plants have not caused illness.

A seaweed called short *ogo,* or *limu manauea (Gracilaria coronopifolia),* caused illness in eight Hawai'i residents in 1994. The suspect seaweed was harvested from the wild in Maui waters. No illnesses have been reported from commercially cultivated plants or from *ogo* sold in grocery stores, pictured here. At *left* is *ogo namasu;* on the *right* is pickled *ogo.* (Susan Scott)

Another seaweed commonly eaten in Hawai'i is *limu kohu (Asparagopsis taxiformis),* a favorite among Hawaiians. This red, Christmas-tree-shaped seaweed grows in shallow areas where waves break. *Limu kohu,* which grows up to 7 inches tall, contains a toxin called bromoform that may cause illness when eaten in large amounts.

Mechanism of Injury

Researchers recently discovered a blue-green alga growing on the surface of *limu manauea.* The alga produces aplysatoxin, a potent marine poison. The cause of this unusual growth is unknown. Researchers found no evidence of pesticide or sewage pollution near the harvest area of Hawai'i's recent *limu manauea* incident.

Limu kohu naturally contains bromoform. This chemical causes DNA damage in bacteria, a predictor of cancer risk in humans. Inhaling bromoform fumes causes lung irritation.

Incidence

In 1994, *limu manauea* caused illness in eight people in one incident. The seaweed came from the shores of Maui.

Similar seaweeds (same genus, different species) have caused serious illness in separate incidents in Japan, Guam, and California. Three victims died in Guam; one in Japan. In Guam, California, and Hawai'i, the seaweeds causing illness had been harmless previously.

In 1995, a visitor was hospitalized in Kaua'i apparently after eating *limu kohu* directly from the reef. Some Hawai'i residents, however, report frequently eating small amounts of equally fresh *limu kohu* with no ill effects.

One California researcher reports that he consistently felt ill while examining a specimen of raw *Asparagopsis taxiformis.* These symptoms were absent after he soaked the seaweed in fresh water, a common practice in traditional preparation.

Prevention

If *limu manauea,* or any other seaweed, causes burning in the mouth, stop eating it, and stop everyone else from eating it. Victims who merely taste toxic *limu manauea* or eat small amounts do not become ill. This toxin is not inactivated by cooking.

Eat *limu kohu* in small amounts. Soaking the seaweed in fresh water at least half a day may make it safer to eat.

Signs and Symptoms

Eating toxic *limu manauea* causes immediate burning in the mouth and throat. Hot chili peppers added to the dish can mask this toxin-induced burning. People who eat a large portion or more than one serving of toxic *limu manauea* suffer nausea, vomiting, burning in the stomach, and diarrhea for about one day.

Limu kohu (Asparagopsis taxiformis) is a reddish, shallow-water seaweed shaped like a Christmas tree. Hawai'i residents who eat *limu kohu* usually eat it in small amounts because of its strong flavor. This seaweed contains bromoform, which can cause illness. (Isabella Abbott)

The Kaua'i victim who ate unsoaked *limu kohu* (amount unknown) suffered severe irritation of the mucus membranes in the mouth, and a swollen tongue. Rarely, *limu kohu* causes stomach irritation.

First Aid

No specific treatment exists for seaweed poisonings. Stop eating any suspect seaweed immediately. Keep others from eating it. Report incidents to the Hawaii Department of Health. Save the suspect seaweed for testing. Anyone seriously ill should see a doctor.

Advanced Medical Treatment

No antidote or clinically useful tests exist for *limu manauea* or *limu kohu* poisoning. Treatment is supportive. For significantly ill patients, treatment with charcoal is reasonable but unproven. Report all cases to the Hawaii Department of Health.

STAPH, STREP, AND GENERAL WOUND CARE

Basic wound care is the same whether you have been cut by a kitchen knife or bitten by a barracuda. With marine wounds, however, the risk of infection is high. Warm ocean water and the mouths and skins of marine animals host various bacteria, including species of *Staphylococcus (Staph)* and *Streptococcus (Strep)*. These bacteria, also common on land and human skin, are the leading cause of marine infections in Hawai'i. Some wounds are complicated by venom, broken teeth, other animal parts, or by bacteria found only in water. For details on specific waterborne bacteria, see *Mycobacterium marinum* and *Vibrio* Infections.

Mechanism of Injury

Most marine wounds are superficial cuts and scrapes, only slightly breaking the skin. Even the slightest abrasion, though, creates an infection risk from bacteria, always present in ocean water and on human and animal skin.

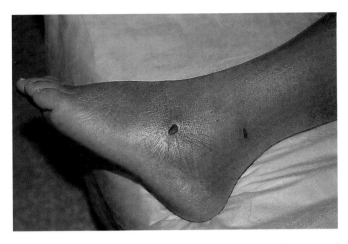

This serious infection on the foot of a North Shore surf photographer began as two minor coral cuts. Cultures revealed *Staphylococcus* and *Streptococcus* species growing in the wound. (Susan Scott)

Although warm, bacteria-laden water soaks all marine wounds, only some wounds become infected. Factors influencing infection rates include the following:

Wound location.   Some areas of the body are more prone to infection than others. Wounds on the scalp and face are the least likely to become infected; wounds on the hands and feet are the most susceptible.

Wound type.   Crush injuries resulting in injured and dead tissue have higher rates of infection than sharp cuts. Deep, large wounds expose more tissue to bacteria and thus increase the risk of infection. Skin flaps on a wound can trap bacteria and create areas that foster infection.

Number of bacteria.   Usually, it takes large numbers of bacteria to infect a wound, but the numbers are not constant. Scalp wounds, for instance, need more bacteria to become infected than hand wounds. Wounds with any object in them (such as a tooth fragment, a piece of coral skeleton, or sutures) need less bacteria inside to start an infection than wounds with nothing in them.

When bacteria in any wound reach high enough numbers to overwhelm the body's natural defense system, an infection sets in, even in healthy people who practice meticulous wound care.

Individual immune systems.   People with damaged immune systems, such as those with liver disease, diabetes, AIDS, or who have had their spleens removed or are taking steroids, have a higher risk of wound infection than others. Also, people with abnormal heart valves have an increased risk of getting a blood-borne infection in the heart after a marine wound.

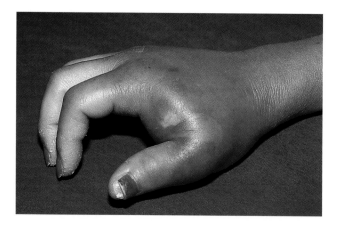

Signs of a wound infection are redness, swelling, pus, warmth, and tenderness. Wound infections need to be treated with antibiotics. Untreated, the infecting bacteria can get into the bloodstream and spread throughout the body, causing life-threatening illness. The serious, potentially life-threatening infection pictured here is cellulitis. (Norm Goldstein, M.D.)

Incidence

Anyone spending time in Hawai'i's ocean waters will eventually get some type of infection, usually a small and superficial *Staph* or *Strep* infection. Other types of bacteria occasionally infect marine wounds. (See Fish Handler's Disease, Seal Finger, *Mycobacterium marinum,* and *Vibrio* Infections.)

Rarely, *Staph, Strep,* or other types of bacteria spread throughout the bloodstream, causing severe tissue damage and, occasionally, death. Even more rarely, a type of *Strep* invades the body's deep tissues and causes an illness popularly called a "flesh eating" infection.

Prevention

Meticulous scrubbing helps reduce the risk of marine infections. See the following First Aid section for information about scrubbing and cleaning techniques.

Signs and Symptoms

Symptoms of most marine infections develop from a few hours to several days after the injury. (*Mycobacterium marinum* is an exception, sometimes appearing months after an injury). Typical signs of infection are redness, swelling, warmth, pus, and pain.

Any size wound may have an object, such as a piece of coral, embedded inside. Suspect this in wounds that heal poorly, hurt out of proportion to the injury, or remain persistently infected. Numbness or inability to move a finger or toe may be a sign of tendon or nerve damage. Infections can spread throughout the body, some (particularly *Vibrio*) extremely quickly. Fever, massive swelling of an area, and collapse are signs of a potentially life-threatening infection.

✚ First Aid

Use the following techniques to decrease the risk of infection in minor marine wounds.

To clean a wound.   Gently pull the edges of the skin open and remove embedded material either by rinsing or with tweezers. Then, rub directly inside the cut with gauze or a clean cloth soaked in clean water. Actively scrub. Rinsing does not remove as many bacteria from wounds as scrubbing.[1] After scrubbing, rinse thoroughly.

Do not delay scrubbing and rinsing for lack of sterile supplies. If only tap water is available, use it. The bacteria count in tap water is extremely low. Never scrub an open wound with ocean water, which often contains large numbers of bacteria.

Antiseptics check the growth of bacteria on, or in, living tissue. One common antiseptic is povidone-iodine solution. It reduces the risk of

A doctor in the ER scrubs a laceration from a surfboard skeg with a dilute povidone-iodine solution. Note that he rubs the sponge directly on the exposed (and numbed) flesh. This is the best method of removing bacteria from a cut. (Craig Thomas, M.D.)

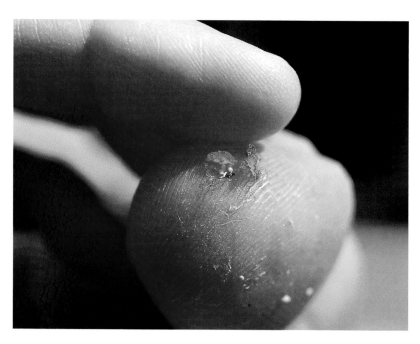

If these sand grains remain in this coral cut, they could start an infection. Also contributing to possible infection are the surrounding skin flaps, which can harbor bacteria. Cut skin flaps off such wounds, then scrub well. (Craig Thomas, M.D., and Susan Scott)

wound infections by slowing the growth of bacteria present in a wound. Full strength povidone-iodine, however, can cause tissue damage.[2] The old-fashion notion that if it burns it must be beneficial is untrue. To make a safe antiseptic solution for scrubbing and rinsing wounds, dilute one part of full-strength povidone-iodine with ten parts of water. (You can buy povidone-iodine already diluted—check the label.) Use these solutions as an addition to wound scrubbing, not as a substitute.

Cut off loose, superficial skin flaps, which create moist, dark areas that can harbor and promote bacteria growth. Do not cut off deep skin or deep tissue flaps. They often grow back together when taped or sutured in place.

Controlling bleeding.   For bleeding wounds, press a clean cloth directly against the wound until bleeding stops. If bleeding persists or the edges of a wound are jagged or gaping, the victim probably needs stitches.

Taping a cut shut is often an appropriate alternative to stitching. Taping, however, may leave a more visible scar than suturing. Wrap tape tightly enough about a wound to close the edges and stop bleeding, but not so tight the wrapping cuts off circulation. Check that fingers or toes near any taped wound remain pink and warm.

To clean a puncture wound.   Puncture wounds differ from open wounds in that dirt, venom, and bacteria may be pushed deep into the flesh during the injury. Since it is difficult to scrub or rinse these substances out of the wounds, punctures are prone to infection. No evidence shows that the folk remedy of squeezing a puncture wound to make it bleed decreases the rate of infection.

Clean and rinse puncture wounds like any other wound, being particularly alert for embedded objects. If you suspect something is inside a wound, see a doctor immediately. Objects left inside often cause infections.

About antibiotics.   Antibiotics are substances that can kill or greatly inhibit the action of bacteria. Most antibiotic ointments are available without a prescription (over-the-counter). Studies show that antibiotic ointments may decrease the rate of infection.[3] Avoid ointments containing neomycin, which frequently causes skin irritation.

Mupirosin, a prescription antibiotic ointment, treats some superficial infections as well as an oral antibiotic.[4] Your doctor may prescribe this for minor skin infections.

Once an infection starts, scrubbing with soap, iodine, hydrogen peroxide, or any other disinfectant is not enough. *Staph* and *Strep,* and all other bacterial infections, need antibiotic treatment.

Advanced infections, especially with fever and chills, can be medical emergencies. Seek help immediately.

All ocean wounds, large and small, carry the risk of tetanus (lockjaw), a deadly bacterial infection. Update your tetanus booster shot approximately every five years if you frequently get marine cuts.[5] If you are not sure about the date of your last tetanus shot, get a booster.

About bandages.   Bandages (dressings) help control bleeding and protect cuts from sand and dirt. Bandages can also hide early signs of infection. Check daily under all bandages for redness and swelling, the first signs of infection.

About scars.   After a scab falls off, scars from most wounds are pink and raised. The scars fade, soften, and shrink for approximately six months. During this time, sun exposure can cause permanent darkening of a scar. Apply waterproof sunscreen daily, 30 SPF or higher, to minimize discoloration.

To minimize scarring, see a doctor for large wounds, objects embedded in wounds, or at the first sign of infection.

Advanced Medical Treatment

Marine infections can be difficult to diagnose and manage. These organisms often require different culturing methods and broader antibiotic coverage than nonmarine infections.

Antibiotic therapy is never a substitute for thorough scrubbing, exploring, irrigating, and debriding of all marine lacerations. Use soft-tissue x-ray, ultrasound, or computed tomography to locate suspected foreign bodies.

In healthy people with minor marine infections, *Staph* or *Strep* are the most likely culprits. Apply mupirocin ointment three times a day for superficial *Staph* and *Strep* skin infections. For localized redness and swelling, add cephalexin, 500 milligrams four times a day. This antibiotic is usually effective and is not photosensitizing. If the infection persists or worsens, or if blisters develop, continue the cephalexin and add doxycycline, 100 mg twice a day, or ciprofloxacin, 750 mg twice a day.[6] This regimen covers the marine organisms most likely to be causing the infection.

Patients with rapidly spreading infection, fever, or generalized illness may need intramuscular or intravenous therapy. Consider ceftriaxone, 2 grams a day with either doxycycline or ciprofloxacin. Do not wait for culture results to begin treatment.

A severe infection with blisters may mean a life-threatening *Vibrio* infection. Purplish swollen tissue surrounding a wound may be fish handler's disease *(Erysipelothrix rhusopathiae)*. Slow-growing nodules suggest *Mycobacterium marinum*. (For more information, see *Vibrio* Infections, Fish Handler's Disease, and *Mycobacterium marinum*.)

Prophylactic antibiotics.   Discuss the risk of infection with patients who have any marine wound. Do not prescribe prophylactic antibiotics for healthy patients with minor injuries and no sign of infection. Antibiotic therapy has minimal effects on infection rates.[7] Also, starting prophylactic antibiotics makes wound infections more difficult to treat when they do occur. In addition, all antibiotics have side effects. One is an increase in sun sensitivity, a potentially dangerous condition in Hawai'i. Common antibiotics associated with photosensitivity are tetracycline, doxycycline, sulfonamides, and ciprofloxacin.[8]

Although using antibiotics to decrease the incidence of wound infection is unproven, early antibiotic therapy is reasonable for patients with large, deep lacerations or those with impaired immune systems. These patients include people with diabetes, liver disease, splenectomy, AIDS, valvular heart disease, or anyone taking steroids. No clinical-outcome data favor a particular antibiotic regimen for decreasing the risk of marine infections.

Tetanus immunization.   Every oceangoer needs to have completed a tetanus immunization series and had a Td (tetanus toxoid/diphtheria vaccine) booster within ten years. To be safe, give Td to anyone with a marine wound who has not had a booster within five years.

Currently, about fifty cases of tetanus are reported each year in the United States. About half these patients received a tetanus booster after their injury and developed tetanus anyway. Tetanus boosters provide immunity for the next injury, not the current one.

Most tetanus cases develop in people who have low levels of tetanus antibodies. For people with inadequate initial immunization, give tetanus immune globulin (TIG) and Td. Although TIG reduces the risk of tetanus acutely, it also impairs the antibody response to Td vaccination.[9] These patients need advice about completing their vaccination series.

Closing wounds.   Traditionally, doctors have believed that closing contaminated wounds increases the risk of infection, and, indeed, deep sutures are associated with increased infection rates. Recent dog and cat bite studies, however, show that skin sutures do not increase the infection rate.[10] Taping or suturing marine wounds to control bleeding, preserve function, or improve appearance is appropriate.

For local anesthesia in closing wounds, use 0.5 percent bupivacaine buffered with 0.1cc of 8.4 percent sodium bicarbonate per 10cc of bupivacaine. The bicarbonate minimizes pain of injection.[11] Bupivacaine anesthetizes as rapidly as lidocaine, but provides much longer pain relief.[12]

SWIMMER'S EAR (OTITIS EXTERNA)

Infection of the outer ear canal (otitis externa) is five times more frequent in swimmers than in nonswimmers, thus the name swimmer's ear for this common ear infection.

The outer ear canal ends at the eardrum, the membrane separating the outer ear from the middle ear. In adults, the outer ear canal is about 1 inch long. Hair follicles and wax protect the outer one-third of the canal. The inner, more delicate, two-thirds of the canal is where moisture and bacteria sometimes become trapped.

Mechanism of Injury

When the warm space at the inner end of the ear canal becomes wet, or the skin is damaged, bacteria multiply and invade the skin. This invasion neutralizes the normal, slightly acidic state of the canal. Bacteria grow even better in this neutral environment, and soon an infection takes hold. Allergies and skin conditions such as psoriasis and eczema can trigger an infection.

Commonly, the itching sensation caused by early ear infections prompts victims to clean the canal with cotton swabs. This simply packs wax into the far end of the ear canal, further trapping moisture and bacteria. Swabbing also damages the delicate inner skin, allowing bacteria to dig in even deeper.

One way to reduce the risk of external ear infections is to blow warm air from a hair dryer into both ears, particularly after swimming. (Craig Thomas, M.D.)

Incidence

In Hawai'i, this type of ear infection is common among swimmers, divers, surfers, and people who clean their ears with cotton swabs.

External ear infections are the reason for 3 to 11 percent of ear, nose, and throat (ENT) office visits on the mainland. Tropical climate, swimming, and using cotton swabs in the ears all increase the incidence of these infections. The incidence of swimmer's ear does not appear to be correlated with fecal bacteria levels in recreational waters.[1]

Prevention

To avoid swimmer's ear, maintain dry, slightly acidic ear canals and keep the skin intact. This can be done in several ways. The easiest is to buy a swimmer's ear solution of boric acid and rubbing alcohol, available at most Hawai'i drug stores. Instilling a couple of drops of the solution in each ear after swimming, especially on hot, humid days, usually prevents swimmer's ear.

Another method is to instill a drop or two of either plain vinegar (acetic acid), or 2 percent to 3 percent boric acid solution, into each ear after swimming. Then dry the ears. Either tip the head sideways, blowing air briefly into each ear with a hairdryer, or instill a few drops of rubbing (isopropyl) alcohol, a drying agent, into each ear.

The premixed squeeze bottles of swimmer's ear solutions are handy but are no more effective than plain vinegar followed by a blast with a hair dryer or rubbing alcohol drops.

Never insert cotton swabs, or anything else, such as ear plugs, into your ears. This just packs wax into the back of the ear canal, almost always causing trouble.

Signs and Symptoms

The first sign of infection is itching, which soon progresses to pain. Pulling gently on the ear exaggerates this pain, as does instilling vinegar, boric acid, or alcohol eardrops.

Severe infections are extremely painful. When the canal swells shut, patients have difficulty hearing on the affected side. In severe cases, redness and swelling are visible on the external ear.

Rarely, an outer ear infection invades deep tissues, causing a condition called malignant otitis externa. This is not a cancer but an invasive infection of the face and deep structures. Symptoms are a severe earache followed by headache, fever, and nausea. This infection is a medical emergency.

 ## First Aid

Prevention is the best medicine. Use drying drops or a hair dryer to dry your ears after swimming. Never insert cotton swabs into your ears, especially after an infection has started.[2]

For early swimmer's ear, try the drops mentioned above. They may sting at first. If the pain worsens, see a doctor.

For extreme pain, decreased hearing, fever, redness, or swelling of the area around the affected ear, see a doctor immediately.

Cotton swabs are a major cause of ear infections. Never, for any reason, insert cotton swabs into your ear canals. (Susan Scott)

Advanced Medical Treatment

Patients who fail first-aid therapy usually respond to antibiotic drops targeted at *Staphylococcus aureus* and *Pseudomonas aeruginosa,* the most common cause of outer ear infections. Drops containing polymyxin B and neomycin provide excellent coverage. These drugs, however, may cause middle or inner ear injury if the eardrum is perforated. In addition, skin often becomes sensitized to neomycin after repeated use. Alternatives are ofloxacin or gentamicin ophthalmic drops. These antibiotics are not marketed as eardrops in the United States. Both these eye preparations have excellent antibacterial coverage and a low incidence of complications.[3] Use two to three drops four times a day and at bedtime.

If the ear canal is swollen nearly shut, gently insert an ear wick of strip gauze to allow drops to penetrate. Because the ear canal has a rich nerve supply, pain medications are essential. For significant ear canal redness or swelling not responding to eardrops, give oral antibiotics such as ciprofloxacin.

Counsel patients on avoiding the use of cotton swabs, and how to dry ears after swimming.

Malignant otitis externa is a rare but life-threatening condition, requiring hospitalization and intravenous antibiotic therapy. Surgical debridement may also be necessary. Treat long-term with ciprofloxacin or imipenem-cilastatin. Ninety percent of malignant otitis externa occurs in diabetic patients.

TURTLE POISONING

Sea turtles are relatives of land tortoises and freshwater turtles that have adapted to life in the ocean. Of the seven sea turtle species, three are regularly found in Hawaiian waters. The Hawaiian green sea turtle, or *honu (Chelonia mydas),* is the most plentiful. Green sea turtles are named for their green body fat, once coveted for soup. Their shells are mottled black, brown, gold, and olive. These seaweed grazers grow 3 to 4 feet long and weigh up to 400 pounds.

Hawksbill turtles, or *'ea (Eretmochelys imbricata),* are the second most prevalent sea turtle in Hawai'i, but are far less common than greens. Hawksbill turtles have lovely shells, for which they have been slaughtered to near extinction. These turtles are smaller than greens; they grow about 2 to 3 feet long and weigh up to 150 pounds. This turtle swims close to the reef, where it pokes its narrow beak into crevices for sponges and other invertebrates.

People sometimes become critically ill, and even die, after eating green sea turtle meat. No one knows why some turtles are toxic and others are not. Hanauma Bay. (D. R. and T. L. Schrichte)

Leatherback sea turtles *(Dermochelys coriacea)* are deep-water turtles that never come ashore in Hawai'i. These enormous turtles grow up to 9 feet long and weigh up to 1,400 pounds. Although all three of these sea turtle species eat jellyfish, leatherbacks do so almost exclusively.

All sea turtles are in danger of extinction. It is illegal to harass or hunt any sea turtle, or to take its eggs.

Green and hawksbill sea turtles have been implicated in fatal poisonings. In ancient Hawai'i, eating hawksbill turtles was forbidden, or *kapu*.

Mechanism of Injury

The substance responsible for turtle poisonings, chelonitoxin, has not been isolated. It may originate from a substance in the turtles' diet. Some apparently healthy looking turtles are toxic; others are not.

People suffering sea turtle poisoning have nervous system impairment; autopsies show massive liver damage. The toxin appears to have no direct action on the heart and does not cause an allergic reaction.

Incidence

No sea turtle poisonings have been reported in Hawai'i. As of 1987, at least 152 people had died from sea turtle poisoning in the Indo-Pacific region. Five of them were infants breast-fed by women who had eaten sea turtles.

Eating hawksbill turtle meat can cause poisoning and death. All sea turtles are protected in Hawai'i, both by state and federal laws. Hanauma Bay. (D. R. and T. L. Schrichte)

Hawksbills were responsible for ninety-six deaths; greens accounted for forty-one deaths. In fifteen deaths, the turtle species was unknown.

Prevention

Never eat any sea turtle. Not only is killing and eating sea turtles illegal and subject to fines, it could be fatal. The toxin is not inactivated by cooking.

Signs and Symptoms

Symptoms develop from hours to days after eating. Nausea, vomiting, rapid heartbeat, pale skin, severe stomach pain, sweating, dizziness, and a sensation of cold in the arms and legs are common. Victims typically have tongue ulcers that become increasingly severe over a few days and persist for months in survivors. Some patients complain of a dry, burning sensation in the lips, mouth, or throat, or all three. Swallowing is difficult.

In severe poisonings, victims progress from lethargy to coma. Convulsions follow, then death.

 # First Aid

For any of the above symptoms after eating any turtle, go directly to an emergency room. Stop anyone else from eating the turtle. Save the head and meat for identification.

Advanced Medical Treatment

Turtle poisoning is life threatening. No antidote or specific therapy exists. Empty the victim's stomach and administer charcoal. Victims may have multisystem organ failure requiring ventilatory and circulatory support.

Report suspected cases to the Hawaii Department of Health.

VIBRIO INFECTIONS

Vibrio is the name of a group of bacteria that can cause serious illness in humans. One type, *Vibrio cholerae,* causes cholera, a life-threatening diarrheal illness. (See Cholera.) Other types, the noncholera *Vibrios,* also can cause diarrhea. Unlike cholera, however, some *Vibrios* may also cause rapid, life-threatening blood-borne infection. *Vibrio vulnificus* is a particularly lethal species.

Vibrio organisms occur naturally in warm seawater and estuaries throughout the world. Researchers recently identified *Vibrio* species on rocks, on marine animals, and in the water surrounding pristine islets 185 miles from Baja California.

Mechanism of Injury

Vibrio bacteria usually enter the bloodstream through the eating of contaminated raw shellfish (about 60 percent of cases) or by infecting an open marine wound (about 30 percent of cases).

The bacteria can multiply rapidly, quickly overwhelming the body's defenses. *Vibrio* organisms cause extensive damage because of their ability to secrete enzymes that dissolve skin, muscle, and connective tissue. Blood-borne infections from this bacterium develop mainly in people with liver disease, excess iron in the body, alcoholism, or damaged immune systems. Some *Vibrios,* such as *Vibrio parahaemolyticus,* secrete a toxin that attacks the lining of the intestine, causing diarrhea.

Incidence

Infections by these bacteria are not common. When they do develop, they can cause serious illness and death. Several ocean-related *Vibrio vulnificus* wound infections have been documented in Hawai'i. The victims survived.

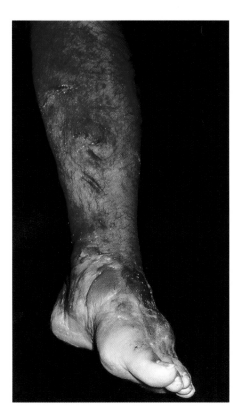

This Hawai'i patient suffered a severe *Vibrio* infection after a barnacle scratch. The person's blood culture was positive for *Vibrio vulnificus.* (C. Samlaska, M.D., Tripler Army Medical Center)

From April 1993 through May 1996, sixteen cases of *Vibrio vulnificus* infection were reported in Los Angeles County. The victims all had eaten raw oysters. Twelve of them had a history of liver disease, and three of these people died.

From 1981 through 1992, seventy-two people in Florida were infected with *Vibrio vulnificus* after eating raw oysters; thirty-six of those people died. All the infected oysters came from restaurants or markets. During those same twelve years, fifty-three Florida people contracted *Vibrio vulnificus* from wound infections or unknown sources. Eight of these infections resulted in death. In *Vibrio vulnificus* wound infections, the victims typically have handled fish or shellfish or received a minor cut in seawater. Most Florida fatalities had liver disease.

In Japan, where people eat large amounts of raw seafood, *Vibrio parahaemolyticus* is a common cause of food-borne diarrheal illness. This infection is rarely diagnosed in the United States, but in 1972, Hawai'i experienced a significant outbreak. Thirty-one people suffered diarrhea after eating raw crabs infected with *Vibrio parahaemolyticus.* All survived.

Other *Vibrios,* such as *Vibrio mimicus* and *Vibrio alginolyticus,* can infect wounds. In 1977, researchers discovered *Vibrio alginolyticus* in superficial wound infections of eight Hawai'i people. Seven of the eight wounds were from marine injuries. All the patients healed with drainage and wound care, with no complications. The relationship of cure to antibiotic therapy in these cases was unclear; 37 percent of the patients received antibiotics to which the bacteria showed resistance.

Prevention

People with damaged immune systems, such as those with diabetes, liver disease, splenectomy, AIDS, valvular heart disease, or anyone taking steroids, should never eat raw shellfish and should avoid contact with seawater when they have open wounds. If they do receive a marine wound, taking prophylactic antibiotics may reduce their infection risk. Meticulous scrubbing of all marine wounds helps reduce the risk of infection. (See *Staph, Strep,* and General Wound Care.)

Commercial oyster growers are currently experimenting with hot and cold treatments to reduce *Vibrio vulnificus* in raw oysters. Cooking oysters is safer than eating them raw.

Signs and Symptoms

In a wound, *Vibrio* organisms can cause a rapidly advancing infection with redness, severe pain, swelling, blistering, and pus. Some of these infections are distinct, bearing large blood-filled blisters that turn into open, draining sores. These infections can spread with remarkable speed, entering the bloodstream and sometimes causing death.

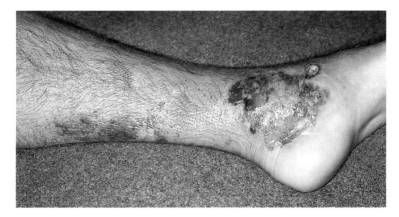

This infection resulted when a sailor landed a tuna near Fanning Atoll and was nicked by the fishhook. Days later, the previously healthy man was critically ill. Navy SEALS were parachuted in to rescue him. The infection was treated with several antibiotics. Although cultures subsequently done at Tripler Army Medical Center were inconclusive, physicians believe the offending organism was *Vibrio vulnificus.* (Scott Norton, M.D., Tripler Army Medical Center)

Vibrio vulnificus organisms are among the most potentially lethal of all marine bacteria. After a person eats these bacteria, they can quickly enter the bloodstream, causing a life-threatening blood-borne infection. This poisoning causes severe flulike symptoms, with fever, vomiting, diarrhea, and chills. Victims develop skin blisters, then black, peeling ulcers, plus inflammation of the muscles. A person at this stage is critically ill.

Initially, it may be impossible to differentiate *Vibrio* wound infections from other types of infections. Suspect *Vibrio* in seriously ill people with marine exposure, especially someone with liver disease. The diarrhea caused by the noncholera *Vibrios* is generally less severe than cholera.

First Aid

For any serious illness after eating raw oysters, or any other raw shellfish, go immediately to an emergency room. *Vibrio* infections, either in a wound or generalized, are life threatening and need antibiotic treatment as described below.

Advanced Medical Treatment

Suspect *Vibrio* in patients with severe, rapidly advancing cellulitis, bullae, and in those with liver disease or damaged immune systems. *Vibrio* wound infections can rapidly progress to necrosis, requiring surgical debridement or amputation.

Many *Vibrio* gastrointestinal infections, especially *Vibrio parahaemolyticus,* cause abdominal cramps and diarrhea, usually less severe than cholera. Fatalities from gastrointestinal *Vibrios* are rare. Wound and gastrointestinal *Vibrio* infections can become systemic, though, causing sepsis, organ failure, and death.

Up to 25 percent of *Vibrio vulnificus*-infected patients develop an explosive, gram-negative sepsis with multisystem involvement. Half of these patients die. Patients who become hypotensive have a mortality rate near 90 percent.[1]

In all suspected *Vibrio* infections, start antibiotics immediately. Do not wait for culture results. Initial therapy must include *Staph* and *Strep* coverage. Use a third-generation cephalosporin, such as ceftazidime, and doxycycline.

For confirmed *Vibrio* infections, two studies recommend doxycycline, 100 milligrams twice a day, or tetracycline, 500 milligrams four times a day.[2] Ciprofloxacin, 750 milligrams two times per day, has also successfully treated *Vibrio* wound infections.[3] Chloramphenicol is an acceptable alternative.

Culture *Vibrio* organisms from blood, wounds, and stool. Use selective culture media with 3 percent salt content. Thiosulfate citrate bile-salts sucrose (TCBS) is a good medium for growing marine *Vibrio* species.

ZOANTHID *(LIMU-MAKE-O-HĀNA)* POISONING

Zoanthids (order Zoanthidae) are anemone-like animals common in shallow tropical and subtropical waters. Zoanthids can be single but usually live in colonies, spreading over rocks like bunches of soft, round flowers about 1/2 inch across. These animals do not build skeletons. In some species, grains of sand embedded in the body walls give support and firmness to these creatures' soft tissues.

Zoanthids are often referred to as soft corals. Like their coral relatives, many zoanthids host symbiotic algae in their cells, which provide oxygen and some nourishment. Some also catch animal plankton on short tentacles bearing nematocysts.

Palythoa toxica (limu-make-o-Hāna). The Hawaiians word for this potentially deadly organism is *limu* (seaweed), but it is a zoanthid, a marine animal related to corals. Zoanthids are sometimes called soft corals. Hawaii Institute of Marine Biology. (William Cooke)

Palythoa tuberculosa is a common zoanthid of Hawai'i and the Pacific. Researchers have found palytoxin in this species in Okinawa. No research has been recorded on Hawai'i specimens. Eniwetok. (William Cooke)

Zoanthid nematocysts do not usually sting human skin. Some species have a powerful toxin, called palytoxin, in their body mucus. At least two of Hawai'i's zoanthids, *Palythoa toxica* and *Zoanthus kealakekuaensis,* carry palytoxin.

Early Hawaiians knew of the toxicity of *Palythoa toxica,* calling it *limu-make-o-Hāna* (the deadly seaweed of Hāna), even though it is not a seaweed. Hawaiians once smeared this toxin on spear tips to make penetration fatal. *Limu-make-o-Hāna* has been reported in surge pools at the Lāna'i Lookout and Blowhole on O'ahu, and in the Hāna area of Maui.

Another of Hawai'i's palytoxin-containing zoanthids, *Zoanthus kealakekuaensis,* lives in the intertidal zone of Kealakekua Bay on the island of Hawai'i.

One cosmopolitan zoanthid species, *Palythoa tuberculosa,* studied in Okinawa, was found to carry palytoxin in its eggs. This species is common in Hawai'i, but no studies have examined the toxin content of local specimens. Hawai'i hosts about seven species of zoanthids.

Mechanism of Injury

Palytoxin is a potent poison that can enter the bloodstream through an abrasion or cut. Because sand is embedded in the body of *Palythoa toxica,* an accidental encounter may abrade skin, allowing palytoxin to enter the body.

Some fish and crabs eat palytoxin, apparently suffering no ill effects. Humans, however, can be severely poisoned from eating marine animals containing palytoxin. This nonprotein toxin causes red blood cells to leak potassium. It also affects muscles by interfering with cells' exchange of sodium and potassium. Muscle breakdown is common.

Palytoxin is a cancer promoter in humans, but conversely, shows anti-cancer activity in laboratory animals.

Palytoxin is probably one of the more deadly members of a family of ciguatera-related toxins. Some researchers believe that some so-called ciguatera deaths are actually caused by palytoxin.

Incidence

A student collector in Hawai'i was hospitalized for two days in 1971 after touching a zoanthid colony to an open wound.

One case of palytoxin fish poisoning was reported in Kaua'i in 1986. The victim ate two smoked mackerel *(Decapterus macrosoma)* imported from the Philippines. He became gravely ill but survived.

In Japan in 1987, two people became acutely ill from palytoxin poisoning after eating parrotfish *(Ypsiscarus ovifrons)*. One of them died. Another person in the Philippines died after eating one-fourth of a crab containing palytoxin. Palytoxin has also been found in a triggerfish called the pinktail durgon, or *humuhumu hi'u kole (Melichthys vidua)*, and in the scribbled filefish, or *loulu (Aluterus scriptus)*.

In an Okinawa study of *Palythoa tuberculosa* eggs, palytoxin showed a lethality far greater than the highly toxic pufferfish eggs.

Prevention

Stay away from anything that looks like clusters of buttons or flat flowers on a rock or in surge zones. Swimmers with open wounds should be particularly careful.

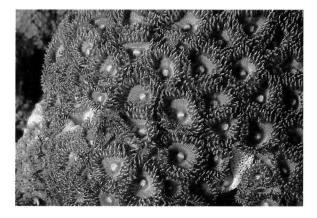

Zoanthus pacificus. No toxin has been associated with this common zoanthid species. Individual zoanthids are only about 1/2 inch across, but their colonies often spread over large areas. Hanauma Bay. (D. R. and T. L. Schrichte)

Palythoa tuberculosa is common on Hawai'i's reefs, spreading like bubbling pancake batter. Although no incidents from this pancake zoanthid have been reported here, and its palytoxin content is extremely variable, its deadly potential calls for caution.

Do not eat fish in the high-risk ciguatera group, which may carry palytoxin. (See Ciguatera Fish Poisoning.)

Signs and Symptoms

Skin contact with zoanthids has been reported to cause swelling and numbness of the area, and a feeling of overall illness.

Palytoxin in the eyes causes inflammation that can last for weeks.

Eating fish or crabs containing palytoxin causes abdominal cramps, nausea, diarrhea, unusual nerve sensations, severe muscle spasms, and respiratory distress. These symptoms can be similar to ciguatera poisoning.

 # First Aid

For skin irritation from a zoanthid, rinse the area copiously with fresh water or seawater. Irrigate eye stings with tap water for at least fifteen minutes.

For eye stings, a feeling of overall illness, or any abdominal pain after eating a crab or fish, go to an emergency room. Stop anyone else from eating suspect seafood. Take the remainder of the food to the emergency room for laboratory identification.

 # Advanced Medical Treatment

For zoanthid contact with cuts or abrasions, scrub and rinse thoroughly. Evaluate for evidence of systemic poisoning.

No specific antidote or clinically useful diagnostic tests exist for palytoxin ingestion. This is a serious poisoning with reported fatalities. This toxin is absorbed within thirty minutes of eating; gastric emptying or charcoal is unlikely to be useful.[1]

Unlike ciguatoxin, palytoxin poisoning has elevated CPK, LDH, SGOT, and urine myoglobin. Determining the difference between palytoxin and ciguatera poisoning is not necessary. Treatment is the same. (See Ciguatera Fish Poisoning.)

No studies of zoanthid eye stings exist. One study of jellyfish eye stings recommends using topical steroids and cycloplegic drops.[2] Eye lesions require ophthalmologic referral to evaluate for corneal lesions and glaucoma.

Part 3

Sports Injuries

With precautions, scuba diving is a relatively safe sport. Because it is a wilderness experience, however, and in a medium alien to the human body, it can never be 100 percent safe. Hanauma Bay. (D. R. and T. L. Schrichte)

DECOMPRESSION ILLNESS

The word "scuba" stands for Self-Contained Underwater Breathing Apparatus. Scuba equipment allows a person to descend underwater while breathing compressed air from a tank. Scuba diving has become an enormously popular sport. The United States hosts an estimated 4 million recreational divers with about 400,000 new divers certified each year.

Most scuba diving deaths are caused by drowning. The next major risk is decompression illness, a collection of related ailments caused by gas bubbles in blood and tissues. The two major decompression illnesses are decompression sickness (the bends) and arterial gas embolism. They differ primarily in the way gas bubbles get trapped in the body.

Air consists primarily of oxygen and nitrogen. In the bends, high pressures dissolve nitrogen in body tissues. A slow ascent allows the nitrogen to be breathed out through the lungs. A rapid ascent, however, reduces pressure too quickly, causing dissolved nitrogen to bubble. The bubbles can lodge in joints (causing victims to "bend"), squeeze the spinal cord, or travel through the veins to the heart.

Normally, bubbles in veins pass harmlessly through the heart into the lungs, where small bubbles are usually filtered out without symptoms. But if a person has a congenital hole in the heart (usually a patent foramen ovale), nitrogen bubbles can bypass the lungs and enter the aorta, the major artery in the body. The bubbles then travel throughout the body, sometimes causing the second serious decompression illness, arterial gas embolus. About 50 percent of arterial gas emboli occur this way.

A congenital hole in the heart, affecting from 10 to 20 percent of the population, is normally not a health problem. In scuba divers it increases the chance of arterial gas embolism by five times.

Another way divers can get an arterial gas embolism is to ascend without exhaling. Pressurized air trapped in the lungs expands, rupturing tiny air sacs. Bubbles then form in the blood and travel to the aorta.

A United States–based international organization called Divers Alert Network (DAN) specializes in diving medical education and emergency consultation for diving accidents. DAN has a 24-hour emergency hotline (919) 684–8111. Any diver interested in the subject of diving medicine can join this organization. For nonmedical, nonemergency information, call (919) 684-2948 weekdays.

Mechanism of Injury

Gas bubbles in the body cause damage by exerting pressure on the spinal cord, joints, or nerves, or by blocking blood flow. The result can be heart attack, stroke, paralysis, and several other conditions, including death, depending on where the bubbles lodge.

Incidence

One hundred divers in fiscal year 1994–1995, and seventy-seven divers in fiscal year 1995–1996, received hyperbaric oxygen therapy for decompression illness at Hawai'i's Hyperbaric Treatment Center (at Kuakini Medical Center in Honolulu). Of these patients, 12 percent suffered from arterial gas embolism. In the United States, 1,163 cases of decompression illness were reported among recreational divers in 1994. Worldwide, approximately one incident is recorded for every five thousand to ten thousand dives.

Incidents of the bends are mainly associated with dives 80 feet and deeper, and repeated dives. Air embolism is much more likely than the bends to cause death in scuba divers. The typical arterial gas embolism incident occurs during a no-decompression dive, within dive table limits, during the first dive of the first day. About 55 percent of the divers make a rapid ascent, a major cause of air embolism.

The average number of recreational diving deaths (from all causes) among Americans over the past ten years is approximately eighty-five per year. In Hawai'i, seven people died in scuba diving accidents in 1993; one death was reported in 1994. Drowning accounts for about 60 percent of scuba deaths throughout the Pacific.

Prevention

Several medical conditions can cause air to be trapped in the lungs. Do not scuba dive if you have asthma, emphysema, or any kind of lung infection.

People with heart abnormalities that allow blood to travel directly from veins to arteries should not scuba dive. Because detecting these abnormalities is sometimes difficult, episodes of decompression illness without an obvious cause may be an indication of this problem. These divers need evaluation by a heart specialist.

Only good swimmers should scuba dive. Drowning is the major cause of death among divers, and good swimming skills may save a life.

Never drink alcohol or take judgment-altering drugs before a dive. Divers who smoke may have an increased risk of suffering an arterial gas embolism.

Recreational divers are at risk for the bends when they approach decompression limits. Scuba divers should stay well within the time limits of any depth, whether using dive computers or dive tables. Never push the limits. Remember that even being conservative with time and depth does not make diving 100 percent safe. Approximately one-half of all bends cases occur within dive table or computer limits.

To make the safest dives, experts currently recommend ascending at a rate of 30 feet per minute. (The traditional rate has been 60 feet per minute.) Going up this slowly can be difficult but is worth the effort. An equivalent way of making a slow ascent is to include a safety stop. A three-minute stop at 20 feet reduces nitrogen bubbles in veins by 50 percent.

Another safety measure is to limit dives to less than 100 feet. Most divers suffering permanent damage from decompression sickness dived deeper than 100 feet.

Never, ever hold your breath while scuba diving. Air embolism can take place in scuba dives as shallow as 4 feet with a breath-holding ascent.

Because air pressure on mountaintops and in pressurized airplanes is lower than at sea level, allow at least twelve hours to pass after a single dive before flying or driving up mountains, such as Haleakalā (Maui) or Mauna Kea (island of Hawai'i). After two or more dives, wait twenty-four hours.

Signs and Symptoms

Symptoms of decompression illness usually appear rapidly, although they may be delayed up to twelve hours or may be triggered by air travel.

Early mild symptoms include unusual fatigue, itching, and joint pain. These may progress to dizziness, visual impairment, chest pain, confusion, weakness, or paralysis. Sudden loss of consciousness, convulsions, breathing difficulty, or paralysis are immediately life threatening.

First Aid

Always regard decompression illness as a medical emergency. The goal is to get the victim to a hyperbaric chamber as quickly as possible. On land, call 911. At sea, contact the U.S. Coast Guard. Air transport should be at as low an altitude as possible.[1] High altitudes risk making a reversible injury permanent.

All divers should be certified in cardiopulmonary resuscitation (CPR). Use it as needed while waiting for emergency medical help. If oxygen is available, give it to the victim at a high flow rate. DAN has a training program and medical kit for administering oxygen to dive accident victims. This is useful when diving in isolated areas.

Never attempt in-water recompression. These victims are at risk for seizures and coma, and are in danger of drowning.

Turn unconscious or vomiting victims onto either side. For convulsions, hold the victim loosely. Never try to force anything into the mouth of a person having a convulsion.

Do not obstruct blood flow to any part of the body by crossing the patient's arms or legs. The victim should not rest his or her head on an arm.

Even if symptoms disappear, it is still essential to contact a doctor from Hawai'i's hyperbaric chamber (at Kuakini Medical Center in Honolulu) for advice.

Advanced Medical Treatment

Get all gas bubble victims to the hyperbaric treatment center as quickly as possible. Do not delay transfer for diagnostic studies.

Decompression illness can be an arterial gas embolism, the bends, or both. Emergency treatment is the same. Initiate life support measures. Give 100 percent oxygen with tight-fitting mask and reservoir, which will improve tissue oxygenation and help decrease the volume of embolized gas. Victims need seizure precautions. Severe cases may need intubation and hyperventilation. Arterial gas embolism victims may have a pneumothorax. Use steroids only for traumatic spinal cord injuries. Steroids are not beneficial in decompression illness.[2]

For air transport, flying at as low an altitude as possible, or in a cabin pressurized to one atmosphere, is crucial. Flying at high altitudes is correlated with residual symptoms after recompression.

Divers are also often at risk for ciguatera fish poisoning, a syndrome including numbness, tingling, and joint pain. Jellyfish stings, too, can cause muscle cramps and a decreased level of consciousness. Both syndromes can mimic decompression illness.

After initial treatment, divers who have arterial gas embolism need echo-cardiography to determine predisposing cause. Patent foramen-ovale and atrial-septal defect can cause recurrent episodes of arterial gas embolism. These people should not dive.[3]

Contact Hawai'i's hyperbaric treatment center (at Kuakini Medical Center in Honolulu) and the DAN 24-hour hotline (919) 684-8111 for medical advice.

DROWNING

In drowning accidents, the time to save a life is at the beach rather than in a hospital. Without oxygen, brain damage results within minutes. Cardiopulmonary resuscitation (CPR) must begin early if the victim is to have a chance at recovery. Victims who have been underwater for less than five minutes and begin breathing on their own within ten minutes of resuscitation usually have good outcomes.

These people at Ala Moana Beach Park understand the risk of children drowning. Five adults watch the children at play in the water. (Susan Scott)

Warm-water victims who require ongoing CPR from the water into the emergency room have a zero rate of survival. Cold-water drownings, not an issue in Hawai'i, are rare exceptions to this dismal prognosis.

Most victims are found floating in or under the water, rather than flailing and screaming for help. Estimates of time-under-water by bystanders, relatives, and even lifeguards are notoriously inaccurate. To be safe, always assume a short submersion time and begin resuscitation.

Mechanism of Injury

Most drowning victims struggle violently in the water before losing consciousness. This exercise leads to faster breathing, heat loss, and a greater chance of inhaling water. Most victims inhale water; a few do not. In either case, breathing stops, preventing the exchange of oxygen and carbon dioxide. The resulting low oxygen content in the body soon leads to cardiac arrest.

Treatments for freshwater and saltwater drownings are the same. Lung injury primarily is related to the amount of fluid inhaled rather than what kind it was.

Incidence

Hawai'i's rate of recreational drowning is the highest in the United States. An average of sixty people drown each year in Hawai'i, about two-thirds of them residents, one-third visitors. Approximately 10 percent of Hawai'i's drownings are children. A study by the City and County of Honolulu Parks and Recreation Department revealed that approximately 70 percent of Hawai'i's children (ages seven through fourteen) cannot swim across a swimming pool.

Drowning is second only to motor vehicle crashes as the most common cause of accidental death for children in the United States. In adults, drowning is the third most common cause of accidental death. Toddlers have the highest drowning rate, followed by teenage males. Of the estimated 500,000 submersion incidents reported in the United States each year, approximately eight thousand result in death.

Alcohol is the chief drug implicated in drownings. An estimated 40 to 70 percent of drowning deaths in American adolescents and adults are alcohol related. An Australian study showed that 64 percent of drowned males had measurable amounts of alcohol in their blood.

Prevention

A responsible adult with good swimming skills should always be present to supervise swimming children. When they come out of the swimming

The cardinal rule in all drowning rescues is to ensure the rescuer's safety. Whenever possible, take a floating object, such as a surfboard, to a drowning victim. Sunset Beach. (Susan Scott)

area, dress children to discourage unsupervised returns to the water. Encourage and support early swimming lessons for children. Many children drown in swimming pools. Fence them.

Good swimming skills are essential for everyone participating in any ocean activity. Everyone, on any kind of boat, should have a U.S. Coast Guard–approved life vest that will float them with head above water even if unconscious. Abstain from drinking any alcohol when going near the water. Even the smallest amount of impaired judgment can mean death in the ocean.

Signs and Symptoms

Conscious submersion victims may have shortness of breath, a cough, wheezing, and chest pain. Unconscious or confused victims may be suffering from lack of oxygen, a head injury, or both.

Chills and shivering are common, even in Hawai'i's warm waters. Water conducts heat away from the body about thirty times faster than air.

The skin of unconscious drowning victims may be bluish and cool. Vomiting is common, especially if the victim has swallowed large amounts of seawater.

✚ First Aid

Of prime importance in drowning incidents is the safety of the rescuers. Whenever possible, throw safety devices to a struggling victim.

If a rescuer must enter the water, take a surfboard or other sturdy floating object along. Multiple rescuers are an advantage.

Withhold no effort in resuscitating an unconscious drowning victim. Use immediate and aggressive measures, even while the person is still in the water. Try mouth-to-mouth ventilation (but not cardiac compressions) while bringing the victim ashore, if the rescuer can remain safe while doing so. When out of the water, place the victim on a firm surface and begin CPR. Continue until the person regains consciousness or an ambulance arrives. If no help is available, discontinue CPR if a person has no pulse after twenty-five minutes.

Tipping a victim upside-down or performing the Heimlich maneuver to drain water from the lungs does no good and only delays resuscitation. Reserve the Heimlich maneuver for patients with an object stuck in the airway.[1]

Take all submersion victims to an emergency room for evaluation.

Advanced Medical Treatment

No unique treatment exists for submersion injuries. Never waste time trying to empty water from the lungs. Assume all victims are hypoxic and initiate oxygen. For unconscious victims, intubate and ventilate with 100 percent oxygen. Use positive end expiratory pressure (PEEP) and suction the airway, all at the scene, if possible.

Submersion victims are at high risk for cervical spine fractures, head injuries, and aspiration pneumonia. When appropriate, transport with cervical spine precautions.

Young victims may have suffered child abuse or neglect.

Submersion victims fall into four groups:

1. Asymptomatic patients (prognosis excellent). In the past, these patients were admitted to the hospital to observe for postimmersion respiratory distress syndrome (secondary drowning). This syndrome is a progression of established pulmonary damage usually detectable in the emergency room within the first six hours. Truly asymptomatic patients with normal pulse oxymetry, vital signs, and chest x-ray can go home after six to eight hours of observation. These patients need detailed follow-up instructions to return to the hospital if they develop any respiratory symptoms. They also may need psychological support for the ordeal of the incident.[2]

2. Symptomatic patients (prognosis good). Coughing, sore throat, and burning in the chest are common symptoms in this stage. Victims who are short of breath are almost always hypoxic. Treat these patients aggressively. Chest x-ray, arterial blood gas, and pulse oxymetry are mandatory. Initial chest x-ray may be normal

in victims with severe pulmonary injury; abnormalities may appear immediately or as late as twenty-four hours postimmersion. Severely hypoxic but alert patients may benefit from mask continuous positive airway pressure (CPAP) or assisted ventilation (BIPAP). Intubate patients who have persistent hypoxia or hypercarbia, altered mental status, or do not improve with CPAP-BIPAP.

3. Comatose patients (prognosis poor). Intubate and hyperventilate to a PCO_2 of approximately 30 millimeters of mercury. Use PEEP to maintain oxygenation. These patients are at high risk for neurologic impairment, multisystem failure, and death. Do not give dextrose empirically. Check blood sugar and treat hypoglycemia. Do not give sodium bicarbonate empirically. Treat acidosis with hyperventilation.

   Steroids are beneficial only for spinal cord injuries. Phenobarbital coma and intracranial pressure monitoring do not improve outcome. Give antibiotics only for specific infections.[3] Calcium channel blockers to reduce brain injury, and pulmonary surfactant replacement, are experimental.

   In comatose children, fixed pupils, male gender, and elevated blood sugar all predict poor outcome.[4]

4. Cardiopulmonary arrest longer than twenty-five minutes (prognosis zero). Nonhypothermic patients who fail resuscitation at the beach do not survive. Multiple centers continue reporting zero survival rates for submersion victims requiring ongoing CPR into the emergency room.[5] Provide psychological support for family and rescuers. Report all drownings to the Hawaii Department of Health.

EAR, SINUS, MASK, AND TOOTH SQUEEZE

During dives, ears and sinuses are susceptible to squeeze or pressure injuries, also known as barotrauma.

Human ears consist of the outer ear canal, the middle ear, and the inner ear. The outer ear canal extends from the external ear to the eardrum, a thin membrane containing multiple, tiny blood vessels. Behind the eardrum is an air-filled space, the middle ear. A passageway called the Eustachian tube connects the middle ear to the nose.

Free divers and scuba divers alike must clear their ears while descending. One method is to blow against pinched nostrils, as shown here. Exhaling into the dive mask prevents mask squeeze. Ningaloo Reef, Australia. (Craig Thomas, M.D.)

Three thin bones connect the eardrum, through the middle ear, to a membrane called the oval window, thus transmitting sound vibrations to the inner ear. This, and another membrane called the round window, separate the middle ear from the fluid-filled inner ear.

Sinuses are enclosed air spaces in the bones of the face. Mucus membranes line the spaces, which, like the middle ear, also have air passages to the nose.

The air spaces of the middle ear and sinuses normally have the same pressure as the outside environment. The tubes connecting the spaces to the nose provide a way to adjust pressures. If the pressure inside does not change when outside pressure does, tissue damage can result.

A similar potential pressure hazard exists in the air space between a diver's mask and eyes. Humans need this air space to see clearly underwater.

Rarely, teeth contain air pockets from cavities or dental work. Underwater pressure can squeeze this air, causing tooth squeeze in divers.

Mechanism of Injury

When a diver descends, water pushes against the eardrum. This pressure makes it stretch and bulge inward, causing the drum to become red, swollen, and painful. If the diver ignores the pain and continues down, serious damage results. Blood vessels of the eardrum and the lining in the middle ear begin to burst at a depth of 3.9 feet. The eardrum can rupture in as little as 4.3 feet of water.

The principle of sinus squeeze is the same as ear squeeze, but sinus squeeze is less common. In sinuses, the air passages to the nose are shorter, making equalization easier. Also, sinuses are more resistant to pressure changes than eardrums.

Mask squeeze occurs when, as a diver descends, water exerts pressure against the air inside the diver's mask. If the diver does not equalize the mask pressure with the outside pressure by breathing out through the nose, capillaries on the surface of the eyes and surrounding tissue can rupture.

Tooth pain results when air pockets from cavities or dental work contract and expand with descent and ascent. This exerts force on nearby nerves, teeth, or bones, causing pain.

Incidence

Ear squeeze, the inability to clear the ears on a dive, is the most common affliction of scuba divers. After a single dive, two-thirds of divers have temporarily injured eardrums. In one study, 100 percent of divers had injured eardrums after eleven days of diving.

Eardrum rupture is rare. Unless the drum ruptures, ear squeeze usually causes only temporary injury.

Ear squeeze is associated with poor underwater visibility, probably because cloudy water makes it difficult to judge changes in depth. Ear squeeze is not associated with a diver's age, sex, or experience. Divers with previous ear and sinus problems (such as childhood ear infections, swimmer's ear, or sinus headaches) have no higher incidence of ear squeeze than those without.

Divers Alert Network reports that one-third of all barotrauma calls are from people with mask squeeze injuries. Approximately six calls a year are about tooth squeeze.

Prevention

Immediately on submerging (at 2 to 3 feet below the surface), blow against your closed lips and nostrils, or swallow, to equalize the pressure on both sides of the eardrum. If this is not done adequately, the Eustachian tube collapses. At this point, attempts to blow air into the middle ear usually are not successful. Ascend to reopen the Eustachian tube, and try blowing again. Never continue down with ear pain. If pain in the ears persists, abort the dive.

Sixty milligrams of pseudoephedrine (an over-the-counter decongestant) taken thirty minutes before diving greatly reduces the incidence and severity of ear squeeze in most divers.[1]

When diving in cloudy water, watch your depth gauge carefully.

Do not dive with a cold or active allergy. These can result in clogged Eustachian tubes or blocked sinus passages.

As soon as a dive mask feels tight against your face, exhale through your nose to equalize pressure.

Do not dive when you have a toothache. Have your dentist fix cavities and replace loose fillings before diving. Wait at least twenty-four hours after dental work to dive.

Signs and Symptoms

The most common symptoms of simple ear squeeze are difficulty clearing the ears during descent, a feeling of pressure in the ear, and ear pain.

Serious pressure injuries to the eardrum begin with discomfort and rapidly increase to severe pain. Actual rupture of the eardrum results in cold water entering the middle ear, causing intense pain, decreased hearing on the injured side, and, sometimes, severe dizziness.

Difficulty in clearing the ears during ascent is uncommon but is a sign of ear damage. Any hearing loss after surfacing requires medical attention.

Sinus squeeze symptoms are facial pain over the injured sinus, sometimes with a nosebleed.

Inner ear trauma is a rare condition caused either by extreme pressure in the middle ear or by the bends of the inner ear. Symptoms are dizziness, hearing loss, and roaring in the ears. These symptoms require immediate medical attention.

In mask squeeze, the whites of the eyes (conjunctiva) become red from ruptured blood vessels. The area around the eyes can become swollen and bruised, also from ruptured vessels. The redness and bruising sometimes worsens after a day or two.

Tooth squeeze causes pain in the affected tooth or in the surrounding area. An air-filled tooth can implode during descent. Sometimes, a filling may dislodge. A cavity can fill with blood.

 # First Aid

Prevention is the key to squeeze injuries. Once they have occurred, stop diving. Take over-the-counter medicine for mild pain. Pseudoephedrine tablets, 60 milligrams, and phenylephrine (Neo-Synephrine) nasal sprays sometimes relieve symptoms of ear or sinus squeeze. For severe pain, hearing loss, balance difficulty, or roaring in the ears, see a doctor.

Mask squeeze with no pain or vision impairment usually disappears over time with no treatment. Simply allow the fluid and blood to reabsorb at its own pace. Symptoms should not appear on subsequent dives if you exhale through your nose as soon as you feel mask pressure against your face.

For tooth squeeze pain, take two 200-milligram tablets of ibuprofen every six hours with food. See a dentist before diving again.

Advanced Medical Treatment

Ear squeeze victims are at high risk for perforated eardrums. These usually heal spontaneously; they occasionally require ENT referral for patching. Although unproven, many practitioners prescribe prophylactic antibiotics for a perforated eardrum. Drops containing both polymyxin-B and neomycin provide excellent coverage. These drugs, however, may cause middle or inner ear injury. In addition, skin often becomes sensitized to neomycin after repeated use. Alternatives are ofloxacin or gentamicin ophthalmic drops. These antibiotics, not marketed as eardrops in the United States, provide excellent antibacterial coverage with a low incidence of complications.[2]

Treatment for sinus squeeze is decongestants, outlined in the First Aid section, above. Use oral antibiotics if chronic sinusitis develops.

Rarely, two kinds of inner ear injuries can occur during dives. They call for opposite therapies. One is inner ear decompression sickness (bends of the inner ear), which requires hyperbaric treatment. The other is rupture of the oval or round windows, often with hemorrhage. In these patients, hyperbaric treatment causes further damage. Both conditions require ENT consultation.

In the past, divers who had suffered any inner ear injury were counseled never to dive again. Research on people who continued diving against medical advice, however, revealed no further injury. This suggests that prohibiting diving in victims of inner ear trauma is unnecessary. These patients need ENT referral for Eustachian tube evaluation, repair, and counseling about various methods of equalizing pressure in the middle ear.[3]

Mask squeeze rarely results in serious injury. Before resuming diving, the patient needs instruction on the importance of exhaling into his or her mask. Consult with an ophthalmologist for mask squeeze patients with severe pain or vision impairment.

Tooth squeeze patients need pain control and dental referral.

Call the Divers Alert Network (DAN) 24-hour hotline (919) 684-8111 for advice about unusual ear injuries after diving. For more information about DAN, see Decompression Illness.

This angler arrived in the emergency room with a fishhook embedded in his thumb. (Cindy Long)

FISHHOOK PUNCTURES

To be effective, fishhooks must easily pierce the skin and tissue of a fish, then stay in place while the fish struggles to get free. The result of these requirements is hooks with extremely sharp tips, equally sharp backward barbs behind the tip, and a curved shape. This design ensures that the harder a fish pulls against the hook, the deeper it bores. The same is true when a fishhook snags an angler.

Mechanism of Injury

Fishhooks can penetrate skin, muscle, and bone. Usually, a hook creates a single puncture wound. If a person pulls on a deeply embedded hook, the backward barb can cause extensive tissue damage.

Incidence

The combination of sharp hooks, slippery fish, rocking boats, and, occasionally, intoxicated anglers, makes fishhook injuries extremely common.

Prevention

Remove a hook from a fish's mouth with a pliers or similar tool. If the hook is deeply embedded in the fish, kill the fish first, then cut the hook out.

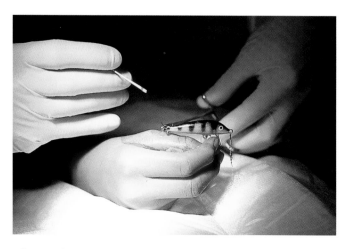

After numbing the area, slide a hollow needle of similar diameter as the fish-hook down the shaft. (Cindy Long)

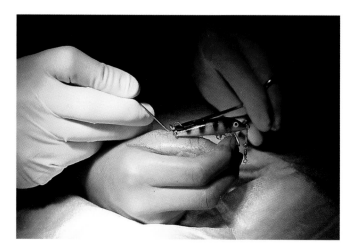

When the needle hits the barb, rotate the needle to cover the barb. (Cindy Long)

If you intend to release a fish with a hook caught deep in its throat, cut the line or the upper part of the hook and leave the rest. Sometimes the remaining portion of the hook will rust and disintegrate, leaving the fish unharmed.

Always wear shoes in fishing boats. Cast with care.

Signs and Symptoms

Fishhook wounds are often puncture wounds with the hook still hanging from the site. Pain is usually minimal unless the hook moves.

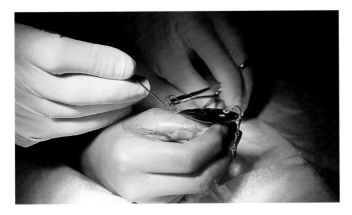

With the barb covered, the hook can easily be backed out of the victim's thumb. (Cindy Long)

 **First Aid**

To remove a hook embedded superficially in the skin, try looping a fishing line around the curve of the hook, then jerk the line parallel to the skin while pressing down on the shaft to disengage the barb.

Never try to yank a deeply embedded, barbed hook from a person's skin or muscle. When possible, cut any lines, bait, or lures from the hook, then hold it as still as possible while going to an emergency room. If the wound is bleeding, apply gentle pressure around the hook's entry site.

If you are far from an emergency room, try removing the hook by either of the methods detailed in the following Advanced Medical Treatment section. The traditional method of advancing the hook through tissue requires a tool to cut the shank off the hook. The other, better technique requires a needle for small hooks or a forceps (or needle-nose pliers) for large ones. Both methods hurt considerably without anesthesia. After removing a fishhook, clean the wound following the techniques in Part 2, *Staph, Strep,* and General Wound Care.

If the hook is in an eye, do not try to remove it and do not press anything against it. Hold the hook as still as possible and rush the victim to an emergency room.

Hooks causing heavy bleeding or numbness, or those that penetrate a joint, should also be held immobile en route to medical help. Pressure around the hook may control bleeding.

 Advanced Medical Care

For routine punctures, clean the area, then infiltrate with buffered bupivacaine to control the pain of removing the hook. To minimize pain, insert the anesthetic needle along the shaft of the hook.

The following method works best for removing fishhooks: Clean the area with a topical antiseptic such as dilute (ten to one) povidone-iodine. For small hooks (after the area is numb), slide an 18-gauge needle into the entrance wound and along the shaft of the hook until you touch the backward-pointing barb. Rotate the bevel of the needle to cover this barb. With the barb of the hook inside the needle, the needle and hook back out easily.

For large hooks, grasp the barb with a splinter forceps or mosquito clamp. Then back out the hook and forceps together.

This technique of covering the barb and backing the hook out rarely fails, and leaves the victim with only one hole in the skin.

The traditional technique for removing hooks is effective but less desirable, since it requires advancing the hook out through the skin, thus creating a second wound in the victim. To use this method, clean around the shaft of the hook and the area that will probably be the exit site. Then anesthetize both areas. When the areas are numb, push the hook with a needle driver or pliers until the hook exits the skin. Cut the shank off the hook: Small shanks can be cut with wire cutters; large shanks require bolt cutters. After cutting the shank, advance the hook out through the exit wound.

After any hook removal, follow the wound treatment details in Part 2, *Staph, Strep,* and General Wound Care.

For complex injuries, such as a hook in an eye or a joint, stabilize the hook and call the appropriate specialist to assist with removal.

SEASICKNESS

Seasickness ruins many an ocean outing in Hawai'i, where winds are strong and seas are rough. Recreational boaters suffer the most, because small boats rise and fall faster than large boats. It is the rising and falling of a boat, appropriately called the heave, that induces seasickness in most people. Pitching, where the front of the boat rocks up and down, causes some additional seasickness, but side-to-side rolling contributes only slightly to the worsening of this malady.

Mechanism of Injury

No one knows the exact cause of this common affliction, but medical researchers agree that seasickness, and motion sickness in general, is the result of sensory confusion in the brain.

Humans, and most animals, are equipped with sensors that constantly provide the brain with information about the environment. These sensors—the eyes, the organs of the inner ears, and structures in the joints, tendons, and muscles—are all linked to the nervous system. They work together to keep the body in balance.

An unfamiliar environment, such as a heaving boat, can cause the sensors to send mixed signals. The eyes, for example, register stability, yet the other sensors register movement. In some people, the mixed messages result in nausea and vomiting.

The exact cause of seasickness appears to involve chemical messengers in the brain. Experiments in blocking or stimulating these messengers (acetycholine and serotonin) can block or induce seasickness. Why some people are more susceptible than others is unclear.

The brain eventually sorts out conflicting sensory information and seasickness ends. The amount of time this takes varies depending on the person, the boat, and the marine conditions.

Some people get sick again after returning to land because the brain has adjusted to motion as the norm. The readjustment time on land is usually a few minutes to a few hours.

Incidence

Nearly 90 percent of humans get seasick. Many cats, dogs, rats, and other mammals do, too. Women tend to get seasick slightly more often than men: One study showed that 55 percent of women and 45 percent of men were seasick during ferry rides.[1]

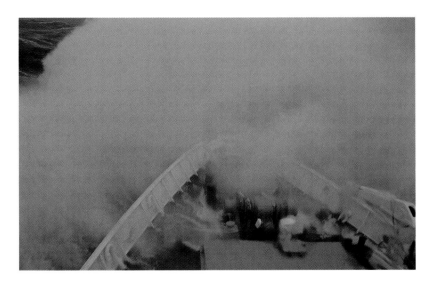

Because a view of the horizon is a stable visual reference, people are less likely to have motion sickness on the bridge of a ship than belowdecks. This view of 60-foot waves in 85-knot winds is from the ship's bridge. Cape Horn. (Craig Thomas, M.D.)

This multiple exposure illustrates the mismatch between vision and motion that causes seasickness. Sometimes, lying down and closing the eyes reduces such sensory confusion and relieves nausea. (Jerry Hughes, M.D.)

Infants under two years old rarely suffer from seasickness. The malady reaches a peak in children from two to twelve years of age. Susceptibility to seasickness progressively declines throughout a person's life, and is less common after fifty years of age.

One theory about why a large percentage of people are susceptible to seasickness is that this is an evolutionary adaptation against poisoning. Many toxins disrupt nerve communication, producing sensory confusion. Vomiting eliminates the toxins.

Prevention

Prevention is everything in seasickness. Most studies show that drugs for seasickness work much better at preventing nausea and vomiting than in stopping these symptoms once they have started. No drugs are a sure cure. The only way to know what works for you is to try them, one at a time. For people who are extremely sensitive to seasickness, stronger prescription drugs may be necessary. For details about over-the-counter and prescription drugs, see First Aid and Advanced Medical Treatment below.

The number of nondrug methods for preventing seasickness is large. In one controlled study, acupressure wrist bands offered no protection against seasickness for either high or low susceptible subjects.[2]

For many people, diesel fumes and other odors promote seasickness. Stay upwind of offensive odors.

Seasickness is usually worse in enclosed, windowless cabins. One study showed that the act of looking out of windows, or watching a horizon, reduced seasickness.[3] Try driving the boat or staring at the horizon.

Hawai'i is notorious for huge, rough seas. To avoid an unpleasant fishing trip or channel crossing, wait until winds are light and seas are calm before embarking.

Signs and Symptoms

Early signs of impending seasickness are stomach awareness, drowsiness, salivation, pallor, cold sweats, headache, general discomfort, depression, anxiety, and a feeling of overall misery. Intense nausea may be followed by vomiting.

 # First Aid

Over-the-counter antihistamines are the first choice for mild to moderate motion sickness. Try meclizine, 25 milligrams every twelve hours; dimenhydramate, 50 milligrams every four to six hours; or cyclizine, 50 milligrams every six hours. Take the medication approximately one hour before you anticipate the motion to start. The long duration of meclizine is an advantage. These seasickness drugs have bothersome side effects. The main complaints are drowsiness, incoordination, impaired decision-making ability, and dry mouth. Ephedrine, 25 milligrams, or pseudoephedrine, 60 milligrams every six hours, added to antihistamines may reduce seasickness more than antihistamines alone but does not reduce the drowsiness caused by antihistamines.

Several recent medical studies suggest that 1-gram tablets of powdered gingerroot helps prevent seasickness. Gingerroot acts directly on the lining of the stomach.[4] Use it alone or with the above drugs. Once a person is seasick, oral drugs do little to help. Even if the victim does not vomit them, pills still do not usually relieve symptoms. Suppositories, skin patches, or injectable drugs, however, often relieve symptoms. (See Advanced Medical Treatment, below.)

Lying down near the center or rear of the boat (where the motion is less severe) and closing the eyes often helps, as does staring at the horizon.

For long bouts of seasickness in hot weather, drink water in small sips. Do not worry about eating. No one dies from seasickness (although many people feel like they are dying). In time, the brain accepts the tilting boat as normal and stops sending nausea and vomiting signals.

Advanced Medical Treatment

No one drug always works for all seasickness. The only sure cure is going ashore.

Prolonged seasickness may cause misery and dehydration but is not life threatening. Antihistamines and sympathomimetic drugs are the best choice for moderately sensitive people. These drugs have fewer side effects than anticholinergic drugs. Scopolamine, an anticholinergic drug, has been widely studied and is often used in transdermal patches. For severe motion sickness, scopolamine appears to be the best choice.[5] Oral and standard patch forms of this drug are not currently available in the United States because of manufacturing problems. The company making this drug states that its transdermal patches will be reintroduced.

Some pharmacists will prepare a transdermal scopolamine gel to be rubbed into the skin. This begins working in one hour and lasts eight to twelve hours. Oral scopolamine, 0.3-milligram tablets, is widely available over-the-counter throughout the rest of the world.

Adding a sympathomimetic drug, such as dextroamphetamine, 5 milligrams every eight hours, improves the antiemetic effect of scopolamine. Reserve this for extremely sensitive patients.[6]

After vomiting begins, treatment should be an antiemetic. Try prochlorperazine or promethazine suppositories or injections. Victims appreciate and welcome the sedation and the antiemetic effects of these drugs.

NASA tested the following drugs and found no clinical effect in treating seasickness: naloxone (Narcan), diazepam (Valium), cannabinoids (marijuana), and dextromethorphan (cough medicine).[7] Ondansetron, a selective serotonin antagonist, is ineffective in motion sickness.[8]

SHALLOW-WATER BLACKOUT

People swimming with a mask and swim fins sometimes dive below the water's surface in a maneuver called free, or breath-hold, diving. In free diving, the person inhales deeply, then swims underwater until the urge to breath forces the diver to the surface.

Usually, free diving is not dangerous. Some people, though, take fast, forced breaths (hyperventilate) before diving to stay underwater longer. This can cause a condition known as shallow-water blackout, in which a person loses consciousness underwater.

Free divers risk blacking out and drowning if they hyperventilate before going down. Hanauma Bay. (D. R. and T. L. Schrichte)

Mechanism of Injury

Rapid, forced breathing causes exhalation of carbon dioxide at a faster rate than normal. Underwater, this can be deadly because the build-up of carbon dioxide in the bloodstream drives the urge to breathe. Therefore, when a free diver hyperventilates and then dives, the body's muscles use up oxygen without the usual accumulation of carbon dioxide in the bloodstream. Without enough carbon dioxide, the diver does not feel the urge to breathe until it is too late.

Shallow-water blackout often occurs just below the surface as the diver ascends. As water pressure decreases on the chest, the oxygen pressure in the lungs drops even further, causing the diver to lose consciousness suddenly.

Incidence

Snorkelers and spearfishermen who free dive are at risk for shallow-water blackout. Statistics are difficult to tally because these cases are recorded as drowning deaths. The usual story is that the free diver was a healthy, strong swimmer found unconscious.

Prompt cardiopulmonary resuscitation (CPR) can save a person's life. A Hawai'i lifeguard at Sunset Beach, O'ahu, demonstrates rescue breathing technique on a CPR mannequin. (Susan Scott)

Prevention

Never hyperventilate before free diving. Rest between dives. Before making a subsequent free dive, breathing should be normal.

Always free dive with a companion. Keep an eye on each other underwater.

Signs and Symptoms

In shallow-water blackout, the diver loses consciousness suddenly, without warning or struggle. After this, the body usually begins breathing automatically, usually drawing water into the lungs.

Above water, hyperventilation can cause lightheadedness, numbness, tingling, anxiety, confusion, and muscle twitching, all from a lowered carbon dioxide level.

First Aid

Treat a suspected shallow-water blackout victim with the same aggressive rescue efforts as any drowning victim. (See Drowning.)

Advanced Medical Treatment

Treat these victims aggressively for drowning. For more information, see Drowning.

Rarely, spontaneous pneumothorax and arterial gas embolism injuries (see Decompression Illness) occur in free divers.[1] These patients probably have lung or vascular abnormalities, putting them at high risk for further episodes. Instruct them to stop free diving and scuba diving.

Free divers are not at risk for decompression sickness (the bends).

Report suspected shallow-water blackout incidents to the Hawaii Department of Health.

SUNBURN

The sun delivers three major kinds of radiation to the earth: About 50 percent is visible (light), 40 percent is infrared (heat), and 10 percent is ultraviolet (UV). The two most prevalent kinds of ultraviolet rays are called UV-A and UV-B, and both of them cause skin damage.

Several factors determine the intensity of UV radiation a person receives. One is time of day. Eighty percent of the sun's rays reach the ground between 9 A.M. and 3 P.M. Other factors affecting UV intensity

Young children are particularly vulnerable to sunburn. Besides having their skin covered with a liberal amount of sunscreen, children should wear hats and sunglasses for all beach, boat, and other ocean outings. Waikīkī. (Susan Scott)

This visitor in Waikīkī became sunburned on his first day in Hawai'i, spoiling his plans for a week out-doors. Once your skin is burned, get out of the sun (and stay out of the sun) until the skin has completely healed. (Susan Scott)

are season of the year, height above sea level, and distance from the equator. Radiation is most intense during the summer, at high altitudes, and at low latitudes.

Another way people can get intense UV radiation is from walking, sitting, or lying on white sand or sidewalks; UV rays reflect off these surfaces. Water also reflects early morning and late afternoon rays. And moist skin absorbs more ultraviolet radiation than dry skin.

A key issue in the amount of UV radiation the earth receives is the ongoing human destruction of the earth's atmospheric ozone layer. Ozone blocks UV radiation.

Ultraviolet radiation from the sun can cause severe skin damage in humans, both short-term as sunburn and long-term as skin cancer, as well as premature aging.

Mechanism of Injury

The skin responds to ultraviolet radiation by increasing its production of melanin, the brown pigment in skin cells. The added melanin, resulting in what is known as a tan, absorbs, reflects and scatters ultraviolet rays, offering some protection from further radiation. Also, the skin's surface layer thickens with prolonged exposure, increasing resistance to UV rays.

When UV radiation exceeds the skin's resistance, it burns, inflaming underlying blood vessels and damaging cells. UV rays also damage chromosomes, which leads to skin cancer.

Incidence

Hawai'i's low latitudes, surrounding ocean, white beaches, and high mountains make sunburn common throughout the Islands. People of all races can get sunburned, but those with lighter skin are more susceptible to burning and to skin cancer. Sun-exposed children are particularly susceptible to long-term skin damage and cancer.

Sun exposure greatly increases the risk of skin cancer and causes dry, wrinkled, discolored skin. About one in seven Americans will get skin cancer. Hawai'i's skin cancer rate is higher than anywhere else in the United States. The deadly skin cancer malignant melanoma (a cancer of the melanin-producing skin cells) develops in about seventy Hawai'i residents yearly. About twenty of them die from it. The risk of getting this deadly skin cancer is increased by intermittent sunburn.

Worldwide, the incidence of melanoma is increasing faster than any other cancer.

Prevention

Regular use of sunscreen that blocks both UV-A and UV-B significantly reduces sunburn and precancerous skin lesions. Apply sunscreen to your face and neck every day, making it part of your morning routine. This is particularly important in children. Childhood burns increase the risk of skin cancer in adulthood.

Without sunscreen, fair-skinned people can burn in about ten minutes on a sunny day in Hawai'i. These people should use a sun-protection factor (SPF) of at least 30. With this protection, they can stay out thirty times ten minutes, or three hundred minutes (five hours) before their skin turns red. Sunscreens with higher SPF numbers offer more hours of protection.

Do not depend on T-shirts for protection. White T-shirts have an SPF of only 6. Special UV-protective clothing is now available. Tanning in white-skinned people provides protection of approximately SPF 4.

Avoid the sun from 9 A.M. to 3 P.M. Beach umbrellas and awnings do not always provide complete protection because ultraviolet rays can penetrate some fabrics. Also, ultraviolet rays reflect off sand, water, and sidewalks.

Wear visors or hats and UV-filtering sunglasses in the sun. Corneas, the eyes' windows, can also become sunburned. Chronic sun exposure can cause cataracts.

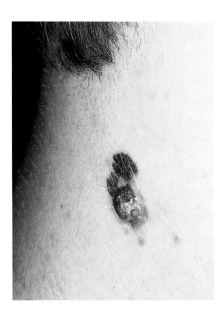

An advanced case of malignant melanoma. This lethal type of skin cancer kills about twenty people a year in Hawai'i. (Ricardo Mandojana, M.D.)

People who spend time in the water should use waterproof sunscreen, applying it liberally and reapplying it after swimming. Adults in swimsuits need approximately one ounce of sunscreen per application. Check the number of ounces per bottle to judge one dose.

Some sunscreens work better than others. For information on the most effective ones, check with Consumers Union, an organization that periodically tests these products and publishes the results in the magazine *Consumer Reports.*

Recent scars are particularly sensitive to sun exposure, often burning and sometimes becoming permanently discolored. Apply waterproof sunscreen, 30 SPF or higher, daily to all scars, especially during the first six months of healing.

Many prescription drugs, including antibiotics such as ciprofloxacin, azithromycin, sulfonamides, and tetracyclines (including doxycycline) increase sun sensitivity. People taking medications should read the package insert or ask a doctor or pharmacist about the drug's sun-sensitive qualities.

Signs and Symptoms

The first and most familiar sign of sunburn is abnormal redness of the skin. This early phase of reddening lasts fifteen to thirty minutes in some people but is not noticeable in others. Delayed redness of the skin begins two to eight hours after sun exposure, reaching a maximum in twenty-four to thirty-six hours. Pain, and sometimes itching, accompanies this redness, commonly called sunburn. This second phase lasts three to five days (given no further exposure) and often results in peeling and itching of the affected areas.

More severe sunburns can result in blisters, swelling, permanent discoloring, or scarring. Burns to large areas of the skin can cause fever, nausea, vomiting, severe headache, and overall weakness.

Some people develop an allergy to sun, resulting in itching and hives each time they are exposed to it.

Effective treatment of skin cancer begins with early detection. Following are the ABCs of melanoma detection. Be suspicious if a skin growth exhibits any of these characteristics.

A. Asymmetry (one half is unlike the other half).

B. Border irregularity (scalloped or poorly defined edges).

C. Color variation within the area (shades of tan and brown, sometimes white, red, or blue).

D. Diameter larger than about 1/4 inch.

Be suspicious of open sores or new lesions that do not heal. These too may be signs of cancer. Dermatologists suggest this reminder for giving yourself an annual check-up: "Examine your birthday suit on your birthday." Have a friend check your back. In one study, people who routinely examined their skin had a 44 percent decreased risk of dying from melanoma.[1]

See a doctor about any unusual changes in your skin.

 # First Aid

For sunburn, get out of the sun immediately and stay out until the burn is completely healed. In animal studies, aloe vera gel, either from the plant's leaves or in commercial preparations, speeds burn healing.[2] Apply the gel to the skin until the burn subsides.

Cool compresses or baths, and products containing the anesthetic benzocaine, can help relieve sunburn pain. Many people become allergic to benzocaine, however, and the ensuing rash only increases a person's misery. Also, benzocaine increases your sun sensitivity.

Ibuprofen, two 200-milligram tablets every six hours (take with food), or other over-the-counter anti-inflammatory drugs, help relieve pain and reduce redness. Applying hydrocortisone lotion or cream to the burned area, besides taking ibuprofen, may reduce redness even further.[3]

None of these drugs, or aloe, prevent long-term skin damage from sunburn.

People with fever, nausea, vomiting, and headache should get out of the sun immediately and see a doctor.

Advanced Medical Treatment

Most burns subside in several days with no treatment. For pain not relieved by ibuprofen or topical steroids, use narcotics.

Patients with hives and swelling may have photo-allergy, a result of interaction between the immune system and UV-A. These patients may benefit from hydroxyzine and oral steroids for several days. Some systemic diseases, such as lupus erythematosus, are worsened by ultraviolet exposure.

Counsel all sunburned patients to avoid further sun exposure, to wear protective clothing, and to use sunscreens. Examining the skin yearly is important in the health maintenance of all Hawai'i residents.

SURF ACCIDENTS

Ocean waves almost always are generated by wind. In the winter, North Pacific storms create large waves that travel across the ocean unimpeded until they reach Hawai'i's north shores. In the summer, South Pacific storms send waves to Hawai'i's south shores. The waves from these distant storms create Hawai'i's famous surf breaks.

The thrill of catching a lift on big waves attracts a diverse group of people using a wide range of wave-riding methods. Some prefer the minimalist bodysurfing approach, using only their body and a pair of swim fins to catch a ride. Others prefer to ride waves on boards, either lying on boogie boards or standing up on surfboards. Still others use the wind, sailing out through the surf on windsurfers, then riding the waves back in.

All methods of surfing can look easy to the casual observer, especially when watching the many experts who live in Hawai'i. It is crucial to remember the following points:

- Hawai'i's surf areas are different from those in other parts of the world. They are seasonally dangerous, have no continental shelf to slow the waves, and often conceal sharp coral reefs.
- Big waves create dangerous currents, called rips, that even the strongest swimmer cannot swim against.
- Each Hawai'i beach presents its own unique hazards.

Mechanism of Injury

Winds make waves in chaotic patterns at storm centers, but as waves travel, they sort themselves into swells, waves that have moved away

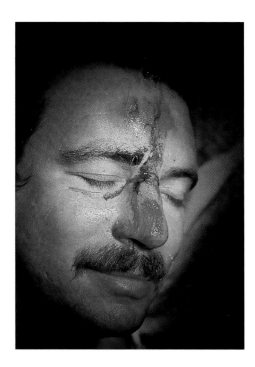

This man suffered extensive lacerations when he hit his face on a coral head while bodysurfing on O'ahu's North Shore. (Susan Scott)

from their place of origin. Larger waves move faster, leaving smaller waves behind. This sorting process is never complete, causing swells to arrive from storms in bunches, called sets. The intervals between sets are called lulls. Sets and lulls are one of the reasons that beaches and coastal areas can be dangerous during high surf conditions. The erratic pattern can surprise even the most seasoned beachgoers.

Hawai'i's surf is known throughout the world for its size, strength, and resulting strong currents. Surfing sports in Hawai'i thus carry the risk of permanent injury or death. Drowning and spinal cord injury are the most common serious events associated with surf.

Each type of surfing has its own specific risk. Bodysurfing has a high risk for spinal cord injuries, resulting when a rider's head gets slammed into the sand. If a person's spinal cord is damaged, permanent paralysis usually results. Board surfers risk lacerations and fractures from surfboards, their own and other surfers'. Because the scalp contains many blood vessels, small cuts there often bleed more than small cuts in other parts of the body. Windsurfers risk foot, ankle, and knee injuries from getting a foot caught in the foot straps. Also, head, neck, and shoulder injuries result from aerial maneuvers and collisions.

Being in the water for long periods of time increase a surfer's chance of getting stung, bitten, cut, or punctured by marine animals. Swimmer's ear, skin ulcers, sunburn, and injuries to knees and elbows are also common.

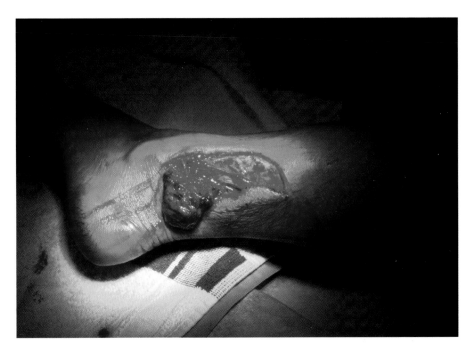

Lacerations from surfboard skegs are a common surfing injury, even in small waves. This North Shore surfer was cut by his skeg in waves 1 to 2 feet high. "If you use this picture," he said, "say it was 20-foot surf." (Susan Scott)

Incidence

In the two-year period 1989 to 1990, sixty-five people in Hawai'i were hospitalized for board surfing injuries, with one death; fifty-five for bodysurfing, with three deaths; and twenty-one for boogie boarding, with no deaths. Because most people with surfing injuries are treated as outpatients, these numbers represent only a fraction of actual injuries incurred.

During an eleven-year Hawai'i study, bodysurfing caused 73 percent of all surfing-related spinal cord injuries. The injuries often result in permanent paralysis. An East Coast study found that of all bodysurfers, middle-aged males were at the highest risk for paralysis.

In 1995, Honolulu City and County lifeguards treated 129 people for neck and back injuries on O'ahu. Sandy Beach had thirty-seven incidents, the highest number of any O'ahu beach. Twenty-one of those incidents were the result of bodysurfing.

A European study found windsurfing (mainly on flat water) safer than tennis and about ten times safer than soccer. In a Hawai'i study listing hospitalizations from surf activities from 1985 through 1988, windsurfing had by far the least admissions.

Windsurfing is evolving, though. Windsurfers today are performing dangerous aerial maneuvers unknown a few years ago. In 1995, at least one windsurfer in Hawai'i suffered a broken neck, and three windsurfing-associated deaths were recorded. One death appeared to have been caused by a heart attack. The other two victims apparently drowned in the surf.

Prevention

Surfers and swimmers need to know the marine "rules of the road" to avoid collisions with boats and should avoid areas with heavy boat traffic. Follow these precautions to avoid injury in the waves:

- Do not surf if you cannot swim.
- Never use drugs or drink alcohol before entering the water.
- Use good judgment. If waves appear to be beyond your abilities, do not go out in them.
- Wear a helmet. A head injury that would be trivial on land can cause death in the water.
- Understand rip currents before entering the water. Most rips are short and narrow. Never swim against a current. Swim to one side, then let the waves carry you back to the beach.
- Accept that waves must be shared. Learn the rules for sharing, then follow them.
- If in doubt, ask lifeguards for advice.
- Bodysurfers should avoid shallow water and big shore breaks.

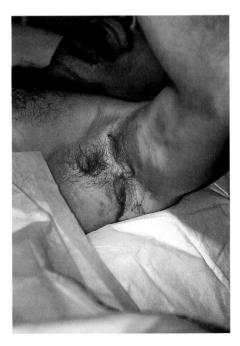

This North Shore surfer was hit in the armpit by the nose of his board. The deep cut required surgical repair. (Susan Scott)

Of all ocean sports in Hawai'i, bodysurfing results in the highest number of victims with a broken neck. Shallow depths at bodysurfing breaks often mean a bodysurfer's head will get slammed into the sand. Sandy Beach, O'ahu. (Craig Thomas, M.D., and Susan Scott)

♦ If you feel yourself going over the falls, lead with your shoulder, not your head.

♦ Surfers and windsurfers should use protective nose guards on their boards.

♦ Windsurfers should tighten foot straps to avoid entrapment.

Signs and Symptoms

Simple scalp lacerations, common in surf accidents, often bleed profusely but are rarely life threatening. Victims with small wounds and little blood loss occasionally have severe reactions to the sight of blood. They may become pale, sweaty, weak, dizzy, and confused. Further injury from a fall after fainting is common in these victims.

People with large blood loss also become weak, dizzy, confused, pale, and sweaty. A victim may be conscious but delirious.

Impacts with a surfboard or reef can cause broken bones, resulting in pain and inability to move an arm or leg. Faintness, difficulty breathing, abdominal pain, or chest pain may be a sign of internal injuries.

Neck injuries are not always obvious. Most neck fractures hurt, but some exhibit minor symptoms. A conscious victim with a broken neck may complain of pain, weakness, difficulty breathing, and tingling or numbness of the arms or legs.

Jellyfish and other marine animal stings can cause intense pain, and sometimes severe reactions, resulting in breathing difficulty. (See Part 1, Box Jellyfish Stings and Portuguese Man-of-War Stings.)

✚ First Aid

For minor cuts, gently pull the edges of the skin open and remove embedded material either by rinsing or using tweezers. Scrub directly inside the cut with gauze or a clean cloth soaked in clean, fresh water. For bleeding wounds, press a clean cloth directly against the wound until the bleeding stops. If bleeding persists or the edges of a wound are jagged or gaping, the victim probably needs stitches. Taping a cut shut is often effective in stopping bleeding but may leave a more visible scar than suturing. For more details on wound care, see Part 2, *Staph, Strep,* and General Wound Care.

Victims who appear pale, sweaty, and nauseated are in danger of fainting. Lower the victim to the ground. Use slings and splints to secure injured arms or legs in position while going to an emergency room. For all serious surf injuries, designate someone to call 911 immediately. Control blood flow by direct pressure whenever possible. In life-threatening cases, use a surfboard leash as a tourniquet while

Jump Rock at Waimea Bay, O'ahu, is notorious as the scene of neck and other injuries. The yellow sign at the top says, "WARNING—DO NOT DIVE OR JUMP." Varied water depth, submerged rocks, sharp coral, slippery rocks, unpredictable waves, high surf, and strong currents can cause serious injury here. (Susan Scott)

Lifeguards at Sunset Beach, O'ahu, demonstrate how they take care of a victim with a possible broken neck. Calling for help immediately is a crucial step in effective first aid. (Susan Scott)

helping the victim to shore. At the beach, control bleeding with pressure, then remove any tourniquet immediately. Administer CPR if necessary. For more details on bleeding, see Part 1, Shark Bites.

After bleeding and breathing are under control, the main concern is a possible broken neck. Assume every unconscious trauma victim has a neck injury until proven otherwise by x-ray. While supporting the victim's head and moving the neck as little as possible remove the victim from the water. Keep the head aligned with the body. Log roll a vomiting victim by turning the head and body together to one side. Watch for breathing difficulties.

If possible, wait for ambulance attendants to transport a victim with a possible neck injury. If ambulance support is not available, hold the person's neck straight and still during transport to an emergency room. This is crucial. Paralysis in a victim with a broken neck may be avoided if rescuers handle the victim correctly.

Advanced Medical Treatment

Surf skegs sometimes break off inside cuts. Sand and other foreign bodies are common. Use soft-tissue x-ray or ultrasound to locate objects in

Neck and Back Injuries for the Year 1995 (Activity)

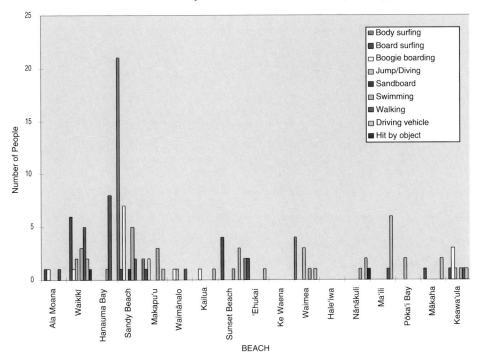

wounds. Remove fiberglass and sand meticulously to prevent infection and foreign body granuloma.

Thoroughly scrub, explore, irrigate, and debride all wounds. Examine for tendon and nerve damage. Repair, or refer, as necessary. Suturing wounds to control bleeding, preserve function, or improve appearance is appropriate.

Do not prescribe antibiotics for minor injuries with no signs of infection, except in immune-compromised patients. For more details about antibiotic therapy, see Part 2, *Staph, Strep,* and General Wound Care.

Some surf injuries can be life threatening. Besides trauma, all surf victims are at risk for near-drowning complications (see Drowning) and marine infections. Bodysurfers are at particular risk for cervical spine injuries.

NOTES AND REFERENCES

Part I   Bites, Cuts, and Stings

Anemone Stings

Notes: 1. Burnett et al. (1994); 2. "Anemones" (1994); 3. Maretic and Russell (1983); 4. Burnett et al. (1994); 5. Glasser et al. (1992); 6. Garcia et al. (1994).

General References: Barnes (1987, 122–127); Clark and Sims (1989, 7); "Coelenterates" (1994); Devaney and Eldredge (1977, 130–147); Fielding (1985, 13–14); Halstead (1987; 1992, 34, 39, 252); Levy (1970); Pearse and Buchsbaum (1987, 163–169); Williamson et al. (in press).

Barracuda *(Kākū)* Bites

General References: Alexander and Proctor (1993, 75–107, 219–243); Michael Beshoner (personal communication, 1996); Clark and Sims (1989, 20); De Sylva (1963); "Fisherman slashed . . ." (1965); Halstead (1992, 12, 243); Hoover (1993, 17–18); Myers (1989, 20); Tinker (1982, 185–186).

Billfish *(A'u)* Wounds

General References: Alexander and Proctor (1993, 75–107; 219–243); Barayuga (1996); Mandojana and Sims (1987); TenBruggencate (1996).

Box Jellyfish Stings

Notes: 1. Ohtaki et al. (1990); 2. Fenner et al. (1993); 3. Exton et al. (1989); 4. Hartwick et al. (1980); 5. Glasser et al. (1992); 6. Burnett and Carlton (1983); 7. Burnett (1994).

General References: "Alum" (1991); Auerbach (1995, 1341–1344); Barnes (1987, 120–121); Burnett (1983; 1992); Burnett et al. (1987, 1990); Clark and Sims (1989, 6); "Coelenterates" (1994); Currie and Wood (1995); "Detergents" (1992); Devaney and Eldredge (1977, 109–111); Halstead (1987; 1992, 34); Hong (1995); Lumley et al. (1988); Mathelier-Fusade and Leyandier (1993); Matsumoto (1995); Meier and White (1995, 105–107); Ohtaki (1990); O'Donnell and Tan (1993); Pearse and Buchsbaum (1987, 160–161); Pongprayoon et al. (1991); Rifkin (1993); Stein (1989); "Sting Aid" (1993); Williamson et al. (in press).

Cone Snail Stings

Notes: 1. "Conotoxins" (1992); 2. Kohn et al. (1960); 3. Valentijn et al. (1992); 4. McIntosh et al. (1993); 5. Kohn (1958).

General References: Barnes (1987, 372, 378); Clark and Sims (1989, 12); Clench and Kondo (1943); Fielding (1985, 47–48); Halstead (1988, 254–255; 1992, 41–43, 247); Hinegardner (1958); Hobson and Chave (1990, 109); Johnson (n.d., 74–79); Kay (1979, 365–383); Kohn (1963); Meier and White (1995, 117–128); Pearse and Buchsbaum (1987, 337); Williamson et al. (in press).

Coral Cuts and Stings

Notes: 1. Hartwick et al.(1980).

General References: Barnes (1987, 127–134); Clark and Sims (1988, 2); Cooper (1981); "Coral" (1992); Devaney and Eldredge (1977, 158–241); Fielding (1985, 18–24); Halstead (1988, 124–125; 1992, 35, 40, 246–247); Hobson and Chave (1990, 101–102); Pearse and Buchsbaum (1987, 170–174); Williamson et al. (in press).

Crab Pinches

General References: Barnes (1987, 603); Clark and Sims (1989, 15); Fielding (1985, 93, 98; 1987, 84); Pearse and Buchsbaum (1987, 492–499).

Crown-of-Thorns Sea Star Stings

Notes: 1. Auerbach (1995, 1346); 2. Meier and White (1995, 132); 3. Shiroma et al. (1994); 4. Edmonds (1995, 139–140).

General References: Barnes (1987, 783); Clark and Sims (1989, 9); Halstead (1988, 189–191; 1992, 46, 247–248); Hobson and Chave (1990, 128); Fielding (1985, 67–68); Pearse and Buchsbaum (1987, 692); Odom and Fischermann (1971); Shiromi (1990); "Starfish Poisoning" (1991); Williamson et al. (in press).

Fireworm Stings

General References: Auerbach (1995, 1351); Barnes (1987, 269–271); Clark and Sims (1989, 14); Devaney and Eldredge (1987, 244–251); Fielding (1985, 35); Halstead (1988, 346–347; 1992, 49–52, 248); Pearse and Buchsbaum (1987, 387–398).

Hydroid Stings

Notes: 1. Chu and Cutress (1955); 2. Rifkin et al. (1993); 3. Meier and White (1995, 107); 4. Glasser et al. (1992).

General References: Auerbach (1995, 1336–1337); Barnes (1987, 101); Clark and Sims (1989, 5); "Coelenterates" (1994); Devaney and Eldredge (1977, 70–105); Fielding (1985, 27, 28; 1987, 60); Halstead (1987; 1992, 31–32, 38); Pearse and Buchsbaum (1987, 136–140); Rifkin (1993); Williamson et al. (in press).

Leatherback *(Lai)* Fish Stings

General References: Halstead (1992, 93); Halstead et al. (1990, 117–118); Hoover (1993, 85); Myers (1989, 132); Tinker (1982, 254); Titcomb (1972, 95); Williamson et al. (in press).

Mantis Shrimp Cuts

General References: Auerbach (1995, 1321); Barnes (1987, 594–595); Clark and Sims (1989, 16); Fielding (1985, 81–82); Pearse and Buchsbaum (1987, 511).

Moray Eel *(Puhi)* Bites

Notes: 1. Randall et al. (1981); 2. Titcomb (1972, 145).

General References: Clark and Sims (1989, 17); Erickson et al. (1992); Halstead (1992, 12, 84, 243); Hobson and Chave (1990, 8–14); Hoover (1993, 50–56); Myers (1989, 44); Randall (1985, 7–9); Tinker (1982, 115–128).

Needlefish *('Aha)* Punctures

Notes: 1. Barss (1985); 2. Bendet et al. (1995).

General References: Alexander and Proctor (1993, 75–107, 219–243); Auerbach (1995, 1320); Clark and Sims (1989, 21); Halstead (1992, 12); Hoover (1993, 91); McCabe (1978); Myers (1989, 71); Tinker (1982, 141–142).

Octopus *(He'e)* Bites

General References: Auerbach (1995, 1353–1355); Clark and Sims (1989, 13); Fielding (1987, 74); Halstead (1988, 305–306; 1992, 43–45, 247); Hobson and Chave (1990, 111); Kay (1979, 588–591); Meier and White (1995, 171–176).

Portuguese Man-of-War *(Pa'imalau)* Stings

Notes: 1. Fenner et al. (1993); 2. Exton et al. (1989); 3. Fenner et al. (1993); 4. Stein (1989); 5. Hartwick et al. (1980); 6. Burnett and Carlton (1983); 7. Glasser et al. (1992); 8. Burnett et al. (September, 1994).

General References: "Alum" (1991); Arnold (1971); Barnes (1987, 110–111); Bengston (1991); Burnett (1983; 1992); Burnett and Carlton (1977); Burnett and Gable (1989); Burnett et al. (1987); Clark and Sims (1989, 4); "Coelenterates" (1994); "Detergents" (1992); Devaney and Eldredge (1977, 105–107); Fielding (1985, 29; 1987, 60); Halstead (1987; 1992, 32–33, 37–38); Kaufman (1992); Lane (1960); Mathelier-Fusade and Leyandier (1993); Meier and White (1995, 107–108); Pearse and Buchsbaum (1987, 150–151); Pongprayoon et al. (1991); "Sting Aid" (1993); Turner (1980); Williamson et al. (in press).

Ray Stings

Notes: 1. Burnett et al. (1986); 2. Fenner (1995).

General References: Auerbach (1995, 1355–1359); Bendt and Auerbach (1991); Clark and Sims (1989, 22); Fenner et al. (1989); "Fish Stings" (1994); Halstead (1992, 11–12, 78–81, 249); Hoover (1993, 104–106) Ikeda (1989); Meier and White (1995, 135–140); Russell (1953); Tinker (1982, 46–56); Williamson et al. (in press).

Scorpionfish *(Nohu)* Stings

Notes: 1. Meier and White (1995, 148); 2. Patel and Wells (1993).

General References: Auerbach (1995, 1359–1362); Clark and Sims (1989, 19); Halstead (1992, 85–89, 249); Hobson and Chave (1990, 23–26); Hoover (1993, 107–111); Kasdan et al. (1987); Kizer et al. (1985); Meier and White (1995, 141–158); Randall (1985, 12–15); Tinker (1982, 419); Williamson et al. (in press).

Sea Snake Bites

Notes: 1. Meier and White (1995, 155).

General References: Auerbach (1995, 1366–1370); Clark and Sims (1989, 25); Halstead (1992, 93–97, 250); McKeown (1978, 61–62); Meier and White (1995, 159–170); "Sea Snakes" (1991); Tu (1987).

Sea Urchin *(Wana)* Punctures

Notes: 1. Auerbach (1995, 1349); 2. Meier and White (1995, 132); 3. Auerbach (1995, 1348); 4. Koerner (1993).

General References: Baden (1987); Baden and Burnett (1977); Barnes (1987, 802–804); Clark and Sims (1989, 3); Fielding (1985, 57–60; 1987, 90–91); Halstead

(1988, 207–208; 1992, 46–49); Hobson and Chave (1990, 122–124); Meier and White (1995, 129–134); Newmeyer (1988); Pearse and Buchsbaum (1987, 714); Russell (1984); Williamson et al. (in press).

Shark *(Manō)* Bites

General References: Alexander and Proctor (1993, 75–107, 219–243); Auerbach (1995, 1309–1317); Balazs (1996); Campbell and Smith (1993); Clark and Sims (1989, 23); Halstead (1992, 5–11); Hobson and Chave (1990, 3–8); Hoover (1993, 112–115); Millington and Wilhelm (1993); Taylor (1993); Tester (1969); Tinker (1982, 1–43).

Sponge Stings

Notes: 1. Sims and Irei (1979).

General References: Auerbach (1995, 1329–1331); Barnes (1987, 71–91); Clark and Sims (1989, 10); Devaney and Eldredge (1977, 53–69); Fielding (1985, 9; 1987, 48, 49); Halstead (1988, 90–92; 1992, 29–31, 245); Meier and White (1995, 85–86); Pearse and Buchsbaum (1987, 71–90); Southcott and Coulter (1971); Yaffee (1970).

Squirrelfish *('Ala'ihi)* Stings

General References: Hoover (1993, 122); Randall (1985, 11); Randall et al. (1990, 60); Meyers (1989, 74, 80).

Stinging Limu Stings

Notes: 1. Sims and Zandee Van Rilland (1981); 2. Tomchik et al. (1993); 3. Burnett (1996).

General References: Auerbach (1995, 1417); Banner (1959); Clark and Sims (1989, 11); Grauer (1959); Izumi and Moore (1987); Magruder and Hunt (1979, 37); "Seaweed Burn" (1959); Serdula et al. (1982); Sims et al. (1993); Wong et al. (1994).

Surgeonfish Cuts

General References: Auerbach (1995, 1365–1366); Clark and Sims (1989, 18); Halstead (1992, 90); Hobson and Chave (1990, 76–85); Hoover (1993, 124–133); Randall (1985, 47–53); Tinker (1982, 373–387).

Part 2   Infections and Poisonings

Allergic Reactions

Notes: 1. Fortenberry et al. (1995).

General References: Moscati et al. (1990); Pollack et al. (1995); Warpinski et al. (1993).

Botulism (from Fresh Fish)

Notes: 1. "Botulism" (1993).

General References: Auerbach (1995, 1405–1406); Centers for Disease Control and Prevention, "Fish Botulism" (1991).

Cholera

General References: Auerbach (1995, 1044–1045, 1401); Barker et al. (1974); Barker and Gangarosa (1974); Barker et al. (1974); Centers for Disease Control and Prevention (1993); "Gastroenteritis—Cholera" (1994); Kontoyiannis (1995); Mintz (1994); Morris (1985); Rheinstein and Klontz (1993); Rippey (1994); Sepulveda et al. (1992); Yamada (1993).

Ciguatera Fish Poisoning

Notes: 1. Palafox et al. (in press).

General References: Auerbach (1995, 1379–1381); Anderson et al. (1983); Cameron and Capra (1993); "Ciguatera Fish Poisoning" DOH (1992); "Ciguatera Fish Poisoning" (1994); Eastaugh (1996); "Fish Poisoning in Hawaii" DOH (1991); Gollop and Pon (1992); Halstead (1992, 147–149, 253–254); "Protocol for the Investigation . . ." DOH (1995); Helfrich (1960); Meier and White (1995, 59–74); Palafox (1988); Shirai et al. (1991); Sims (1987); Swift and Swift (1993); Zlotnick (1995); Williamson et al. (in press).

Fish Handler's Disease *(Erysipelothrix rhusopathiae)*

General References: Auerbach (1995, 1422); Dierauf (1990, 165).

Fishworm Infections (Anisakiasis)

Notes: 1. Deardorff et al. (1986, 1991); 2. Ibid.; 3. Doury (1990).

General References: Ambrose (1985); Auerbach (1995, 1402–1403); Hiramoto and Tokeshi (1991); Pearse and Buchsbaum (1987, 288); Shirahama et al. (1992).

Hallucinatory Fish Poisoning

Notes: 1. Helfrich and Banner (1960); 2. Ibid.

General References: Auerbach (1995, 1378); Banner (1973); Halstead (1992, 158–159); Helfrich (1963); Helfrich and Banner (1960); Hobson and Chave (1990, 38); Sims (1987); Titcomb (1972, 160).

Mycobacterium marinum

Notes: 1. Edelstein (1994); 2. Ibid.

General References: Auerbach (1995, 1421–1422); Clark and Sims (1989, 24); Dierauf (1990, 165–166); Harth et al. (1994); Iredell et al. (1992); Johnston and Izumi (1987); Parent et al. (1995); Paul and Gulick (1993); Thompson et al. (1993); Walker et al. (1962).

Pufferfish Poisoning (Tetrodotoxin)

Notes: 1. Auerbach (1995, 1388).

General References: Auerbach (1995, 1386–1388); Halstead (1990, 159–164, 254–255); Hobson and Chave (1990, 90–94); Lange (1990); Meier and White (1995, 75–84); Sims and Ostman (1986); "Tetrodotoxin" (1992); Williamson et al. (in press).

Sardine Poisoning (Clupeotoxin)

General References: Auerbach (1995, 1385); Girard et al. (1979); Halstead (1992, 148, 156–157); Mandojana and Sims (1987); Melton et al. (1984).

Scombroid Poisoning:

Notes: 1. Auerbach (1990).

General References: Auerbach (1990; 1995, 1385–1386); Halstead (1992, 254); Morrow et al. (1991); Nakata (1994); "Scombroid Fish Poisoning" (1994).

Seal Finger

Notes: 1. Denoncourt (1991); Eadie et al. (1990).

General References: Ching (1994, 4, 7, 10, 22, 28); Dierauf (1990, 165); Markham and Polk (1979); Mass et al. (1981); Rodahl (1943); Sargent (1980), Skinner (1957); Stadtlander and Madoff (1994); Thompson et al. (1993).

Seaweed Poisoning

General References: Abbott (1984, 6, 28–29); Centers for Disease Control and Prevention (1995); Chu et al. (1982); LeCurieux et al. (1995); Magruder and Hunt (1979, 71–73); Marshall (1995).

Staph, Strep, and General Wound Care

Notes: 1. Howell, et al. (March 1993; December 1993); Dire et al. (1990); 2. Bukata (1995); 3. Dire et al. (1995); 4. Britton et al. (1990); 5. Centers for Disease Control and Prevention, "Diphtheria, Tetanus, and Pertussis" (1991); "Tetanus Toxoid" (1994); 6. Millington and Wilhelm (1993); Meadors and Pankey (1990); 7. Cummings et al. (1994); 8. Sanford et al. (1996, 111); 9. Dal-Re et al. (1995); 10. Maimaris et al. (1988); 11. Coventry et al. (1989); 12. Fariss et al. (1987).

General References: Auerbach (1995, 933); Chisholm (1989); Chrintz et al. (1989); Cummings (1995); Dire et al. (1994); Lineaweaver et al. (1985); Rosen (1985); Tong (1996).

Swimmer's Ear (Otitis Externa)

Notes: 1. Calderon and Mood (1982); 2. Krause (1992); 3. Barlow (1995).

General References: Auerbach (1995, 1425); "Otitis Externa" (1992); Strauss and Dierker, 1987).

Turtle Poisoning

General References: Halstead (1992, 174–176); Limpus (1987); McKeown (1978, 66–75); Silas and Fernando (1984); Williamson et al. (in press).

Vibrio Infections

Notes: 1. Hlady et al. (1993); 2. Ibid.; Limpert and Peacock (1988); 3. Meadors and Pankey (1990).

General References: Auerbach (1995, 1034, 1044–1045, 1400–1401); Centers for Disease Control (1996); Johnston et al. (1985); Klontz et al. (1993); Millington and Wilhelm (1993); Pien et al. (1977); "Preventing Serious Raw Oyster-Associated . . ." FDA (1995); Rippey (1994); Rheinstein and Klontz (1993); Sanford et al. (1996); Sims et al. (1993); Tacket et al. (1984).

Zoanthid *(Limu-make-o-Hāna)* Poisoning

Notes: 1. "Palytoxin" (1994) 2. Glasser et al. (1992).

General References: Barnes (1987, 127–128); Clark and Sims (1989, 8); Devaney and Eldredge (1977, 148–157); Fielding (1985, 15–16; 1987, 62); Halstead (1992, 129–130); Hashimoto et al. (1969); Kimura and Hashimoto (1973); Kimura and Yamazato (1972); Kodama et al. (1989); Mandojana and Sims (1987); Pearse and Buchsbaum (1987, 175).

Part 3   Sports Injuries

Decompression Illness

Notes: 1. Almon (1995); 2. Francis and Dutka (1989); 3. Moon et al. (1995).

General References: Adams (1993); Aharon-Peretz et al. (1993); Auerbach (1995, 1176–1208); Bennett and Elliot (1993, 253–266, 376–432); Damron (1992); "Decompression Sickness" (1993); Dovenbarger (1995); Dovenbarger and Uguccioni (1995); Isakov et al. (1996); "1993 Report on Diving Accidents . . ." (1995); Tabrah et al. (1994); Schwartz (1996); "Underwater Diving Accident Manual" (1994); Vann et al. (1993).

Drowning

Notes: 1. Quan (1993); 2. Auerbach (1995, 1221); "Near Drowning" (1994); 3. Ibid.; 4. Graf et al. (1995); 5. Quan (1993).

General References: American Red Cross (1991); Anderson et al. (1991); Auerbach (1995, 1209–1233); Graf et al. (1995); Nichter et al. (1989); Odom (1984); Pratt and Haynes (1986); Quan and Kinder (1992); Quan et al. (1990); Wong and McNamara (1984); Yamamoto et al. (1992).

Ear, Sinus, Mask, and Tooth Squeeze

Notes: 1. Brown et al. (1992); 2. Barlow et al. (1995); 3. Parell and Becker (1993).

General References: Auerbach (1995, 1183–1184), Bennett and Elliot (1993, 267–300); Green et al. (1993); Koriwchak and Werkhaven (1994); "Underwater Diving Accident Manual" (1994, 34–35); Willingham and Dovenbarger (1996).

Fishhook Punctures

General References: Auerbach (1995, 342–343, 361–362); "To Get a Fisherman Off the Hook" (1992); Yardon (1995).

Seasickness

Notes: 1. Lawther and Griffin (1988); 2. Warwick-Evans et al. (1991); 3. Rolnick and Bles (1989); 4. Mowrey and Clayson (1982); Grontved and Hentzer (1986); Grontved et al. (1988); 5. Pingree and Pethybridge (1994); 6. Schmedtje et al. (1988) 7. Lucot and Crampton (1991); 8. "Ondansetron" (1995).

General References: Auerbach (1995, 291–311); Coughenower (1994); Doweck et al. (1994); Gordon et al. 1994); Pingree (1989, 1994).

Shallow-Water Blackout

Notes: 1. Kol et al. (1993).

General References: Auerbach (1995, 1216); Bennett and Elliot (1993, 253).

Sunburn

Notes: 1. Berwick et al. (1996); 2. Rodrigue-Bigas et al. (1988); 3. Visuthikosol et al. (1995); 3. Hughes et al. (1992).

General References: "Burns" (1992); Davis et al. (1989); Goldstein (1993); Kaplan (1992); King and Newcomer (1982); Miller and Bogle (1993); "Putting Sunscreens to the Test" (1995); Westerdahl et al. (1994).

Surf Accidents

General References: Allen et al. (1977); "Causes and Consequences . . ." (1994); Chang and McDanal (1980); Cheng et al. (1992); Hartung et al. (1990); Hartung and Goebert (unpub. ms.); Mettler and Biener (1991); Pacelli (1990); "Preventing Childhood Injury" (1995).

GLOSSARY

Anaphylaxis. Life-threatening allergic reaction.

Anticholinergic drug. A drug interfering with the action of acetylcholine, a chemical involved with the transmission of nerve impulses.

Antiemetic. An agent that stops vomiting.

Antihelminthic. An agent that kills parasitic worms.

Arthrodesis. The surgical immobilization of a joint so the bones will grow solidly together.

Arrhythmia. Alteration of the heart rhythm.

Ascending paralysis. Paralysis beginning at the lower part of the body, then moving upward.

Aspiration pneumonia. Inflammation of the lungs due to the entrance of foreign material such as food particles or ocean water.

Bradycardia. Slow heartbeat.

Bronchospasm. Constriction of the larger air passages of the lungs.

Bullae. Large, thin-walled blisters.

Cellulitis. An acute inflammation of tissues, usually associated with infection.

Charcoal. Absorbs toxins. Used in emergency rooms as an all-purpose antidote. Victims of poisonings either drink it or have it administered through a tube leading from the nose or mouth to the stomach.

Chromosomes. The structures in cells containing genetic information.

Cnidarians. The name of a group of animals that includes sea anemones, hydroids, corals, and jellyfish. All members of this group bear stinging tentacles. Cnidarians were formerly called coelenterates.

Coagulopathy. A blood-clotting disorder.

Coelenterates. Former name of cnidarians.

Conduction delay. Interference with the electrical conductivity of the heart.

CPAP. Continuous Positive Airway Pressure. A method of assisted breathing with a mask, for victims of lung tissue damage.

Darkfield microscopy. A laboratory technique used to identify certain bacteria.

Debride. To remove foreign material, and contaminated or dead tissue, from a wound.

Descending paralysis. Paralysis beginning at the head and face, then moving down the body.

Dive computer. Portable depth gauge and computer, which performs time and depth calculations to minimize the risk of decompression illness.

Dive tables. A standardized table of times at various depths, used to minimize the risk of decompression illness.

Embolized gas. Bubbles in the bloodstream.

Emesis. Vomit.

ENT. An ear, nose, and throat specialist.

Endocarditis. Inflammation of the heart lining.

Endoscope. An instrument used to inspect the inside of the stomach. The endoscope is inserted through the mouth.

Eosinophilia. The increase of eosinophils, a type of white blood cell, in the bloodstream. Commonly associated with chronic allergies.

Epinephrine. A hormone secreted by the adrenal glands in response to stress. Also known as adrenaline.

Erythema multiforme. Rash with multiple red, irregularly shaped lesions.

Foramen ovale. A hole between the right and left sides of the heart that usually closes at birth.

Granuloma. A mass of chronically inflamed tissue usually associated with an infection.

Guillian-Barré syndrome. An acute inflammation of the body's nerves, resulting in temporary paralysis.

Heat labile. Destroyed by heat.

Hepatic failure. Liver failure.

Hypercarbia. Increased levels of carbon dioxide in the blood.

Hypothermic. Below normal body temperature.

Hypotension. Low blood pressure.

Hypoxia. Low levels of oxygen in the blood.

Immune compromised. Any number of conditions in which the body cannot fight infection normally.

Intubation. Insertion of a breathing tube into the trachea, the airway to the lungs.

Lavage. The rinsing out of the stomach through a tube inserted through the nose or mouth.

Leptospirosis. An illness caused by a bacterium carried in the urine of some animals. In Hawai'i, leptospirosis is most commonly transmitted to humans by rats urinating in freshwater streams.

Local reaction. A reaction restricted to one area. Also called a "localized" reaction.

Lupus erythematosus. A chronic disease affecting the skin and often other organs.

Meningoencephalitis. Inflammation of the linings of the brain and spinal cord.

Myasthenia gravis. A chronic nerve and muscle disorder. Usually first noticed in the eyelids.

Necrosis. Tissue death.

Parenteral. By injection.

PEEP. Positive End Expiratory Pressure. A method of assisted breathing through a tube for victims of lung tissue damage.

Peripheral nerve deficit. A malfunction of nerves controlling movement and sensation.

Peritonitis. Inflammation of the lining of the abdomen, usually due to infection.

Pneumothorax. Collapsed lung caused by air leaking into the space between the lung and the chest wall.

Polyneuritis. Inflammation of multiple nerves.

Prognosis. Outcome.

Pulse oxymetry. The measuring of the amount of oxygen in the bloodstream by means of a light shining through a finger or toe to a light detector.

Rhabdomyolysis. The disintegration of muscle.

Self-limiting. Conditions in which the body recovers by itself. This is said of illnesses that run definite, limited courses not requiring medical therapy.

Sympathomimetic. Mimicking the effects of nerve impulses in certain nerves controlling involuntary actions, such as the "flight or fight" response.

Sensitize. Exposure to a substance causes some people to develop antibodies to that substance. This person is said to be sensitized to the substance. At a later exposure, the antibodies can cause severe illness.

Sepsis. Bacteria or viruses in the bloodstream. Often called blood poisoning.

Septic arthritis. Inflammation of a joint due to infection.

Systemic. Affecting the body as a whole.

Tensilon test. A drug test that temporarily improves strength in myasthenia gravis, a chronic nerve and muscle disorder.

Tetanus. Also known as lockjaw. An acute, often fatal, disease caused by the bacterium *Clostridium tetani*. Usually enters the body through wounds.

Thoracic aorta. The major artery of the body, leading from the heart to the abdomen.

Topical. Affecting or applied (like a cream) to the skin.

Transdermal. Absorption through the skin.

Vascular. Pertaining to veins and arteries.

Ventilatory support. Assisted breathing.

Vesicles. Multiple, small blisters.

BIBLIOGRAPHY

Abbott, Isabella Aiona. 1984. *Limu: An Ethnobotanical Study of Some Hawaiian Seaweeds.* Lāwai, Kauaʻi: Pacific Tropical Botanical Garden.

Adams, Mary Jane. 1993. "An Outbreak of Ciguatera Poisoning in a Group of Scuba Divers." *Journal of Wilderness Medicine* 4 (3): 304–311.

Aharon-Peretz, Judith, et al. 1993. "Spinal Cord Decompression Sickness in Sport Diving." *Archives of Neurology* 50: 753–756.

Alexander, Raymond H., and Herbert J. Proctor. 1993. *Advanced Trauma Life Support, Program for Physicians.* Chicago: American College of Surgeons.

Allen, Robert H., et al. 1977. "Surfing Injuries at Waikiki." *Journal of the American Medical Association* 237 (14 February): 668–670.

Almon, Anthony K. 1995. "Cracking the DCS Code." *Alert Diver* (July–August): 36–40.

"Alum." Revised September 1991. Poisindex® Toxicologic Managements 84. 1974–1995. Database online. Denver: Micromedex Inc.

Ambrose, Jeanne. 1985. "Favorite Island Fish May Carry Parasite." *Honolulu Star-Bulletin,* 13 February, p. 1-A.

American Red Cross, Hawaiʻi State Chapter, Keiki Swim Program. 1991. Letter to Hawaii State Legislature.

Anderson, Bruce S., et al. 1983. "The Epidemiology of Ciguatera Fish Poisoning in Hawaii, 1975–1981." *Hawaii Medical Journal* 42 (October): 326–334.

Anderson, K. C., et al. 1991. "Submersion Incidents: A Review of 39 Cases and Development of the Submersion Outcome Score." *Journal of Wilderness Medicine* 2: 27–36.

"Anemones." Revised in October 1994. Poisindex® Toxicologic Managements 84. Database online. Denver: Micromedex Inc., 1974–1995.

Arnold, Harry L., Jr. 1971. "Portuguese Man-of-War ('Bluebottle') Stings: Treatment with Papain." *Straub Clinic Proceedings* 37 (January–March): 30–33.

Auerbach, P. S. 1990. "Persistent Headache Associated with Scombroid Poisoning: Resolution with Oral Cimetidine." *Journal of Wilderness Medicine* 1: 279–283.

Auerbach, Paul S., ed. 1995. *Wilderness Medicine: Management of Wilderness and Environmental Emergencies.* 3rd ed. St. Louis: Mosby.

Baden, Howard P. 1987. "Injuries from Sea Urchins." *Clinics in Dermatology* 5 (July–September): 112–117.

Baden, Howard P., and Joseph W. Burnett. 1977. "Injuries from Sea Urchins." *Southern Medical Journal* 70 (April): 459–460.

Balazs, George. 1996. "Annotated List of Shark Attacks in the Hawaiian Islands." National Marine Fisheries Service, NOAA, Honolulu Laboratory.

Banner, Albert H. 1959. "A Dermatitis-Producing Alga in Hawaii." *Hawaii Medical Journal* 19 (September–October): 35–36.

———— 1973. "Hallucinatory Mullet Poisoning: A Case from Oahu." *Hawaii Medical Journal* 32 (September–October): 330–331.

Barayuga, Debra. 1996. "Danger of Billfishing Claims Fisherman." *Honolulu Star-Bulletin,* 15 March, p. A-2.

Barker, William H., Jr. 1974. "*Vibrio parahaemolyticus* Outbreaks in the United States." *Lancet,* 30 March, pp. 551–554.

Barker, William H., Jr., and Eugene J. Gangarosa. 1974. "Food Poisoning Due to *Vibrio Parahaemolyticus.*" *Annual Review of Medicine* 25: 75–81.

Barker, William H., et al. 1974. "*Vibrio Parahaemolyticus* Gastroenteritis Outbreak in Covington, Louisiana, in August 1972." *American Journal of Epidemiology* 100 (October): 316–323.

Barlow, D. W., et al. 1995. "Ototoxicity of Topical Otomicrobial Agents." *Acta Otolaryngology* 115 (March): 231.

Barnes, Robert D. 1987. *Invertebrate Zoology.* 5th ed. Philadelphia: Saunders College Publishing.

Barss, P. G. 1985. "Penetrating Wounds Caused by Needlefish in Oceania." *Medical Journal of Australia* 143 (9–23 December): 617–622.

Bendet, Erez, et al. 1995. "Penetrating Cervical Injury Caused by a Needlefish." *Annals of Otology, Rhinology and Laryngology* 104 (March): 248–250.

Bendt, R. R., and P. S. Auerbach. 1991. "Foreign Body Reaction Following Stingray Envenomation." *Journal of Wilderness Medicine* 2: 298–303.

Bengston, Kenneth, et al. 1991. "Sudden Death in a Child Following Jellyfish Envenomation by *Chiropsalmus quadrumanus.*" *Journal of the American Medical Association* 266 (11 September): 1404–1406.

Bennett, Peter B., and David H. Elliot, eds. 1993. *The Physiology and Medicine of Diving.* 4th ed. London: W. B. Saunders.

Berwick, Marianne, et al. 1996. "Skin Self-Examination." *Journal of the National Cancer Institute* 88 (January): 17–23.

Beshoner, Michael. 1996. Barracuda attacks. Emergency Department, Kona Community Hospital, Kona, Hawai'i. Personal communication, April.

"Botulism." Revised in December 1993. Poisindex® Toxicologic Managements 84 Database online. Denver: Micromedex Inc., 1974–1995.

Britton, J. W., et al. 1990. "Comparison of Mupirocin and Erythromycin in the Treatment of Impetigo." *Journal of Pediatrics* 117 (November): 827.

Brown, M., et al. 1992. "Pseudoephedrine for the Prevention of Barotitis Media: A Controlled Clinical Trial in Underwater Divers." *Annals of Emergency Medicine* 21: 849–852.

Bukata, Richard W. 1995. "Limiting the Likelihood of Wound Infections." *Emergency Medicine and Acute Care Essays* 19 (October) (unpaginated).

Burnett, Joseph W. 1983. "First Aid for Jellyfish Envenomation." *Southern Medical Journal* 76 (July): 870–872.

———. 1992. "Human Injuries Following Jellyfish Stings." *Maryland Medical Journal* 41 (June): 509–513.

———. 1994. "Mononeuritis Multiplex after Coelenterate Sting." *Medical Journal of Australia* 161 (September): 320–322.

———. 1996. *Jellyfish Newsletter* 15 (July): 7. (Published by International Consortium for Jellyfish Stings, Baltimore.)

Burnett, Joseph W., and Gary J. Carlton. 1983. "Response of the Box-Jellyfish *(Chironex fleckeri)* Cardiotoxin to Intravenous Administration of Verapamil." *Medical Journal of Australia* 2: 192–194.

———. 1977. "Review Article: The Chemistry and Toxicology of Some Venomous Pelagic Coelenterates." *Toxicon* 15: 177–196.

Burnett, Joseph W., and W. D. Gable. 1989. "Fatal Jellyfish Envenomation by the Portuguese Man-of-War." *Toxicon* 27 (7): 823–824.

Burnett, Joseph W., et al. 1987. "Local and Systemic Reactions from Jellyfish Stings." *Clinics in Dermatology* 5 (July–September): 14–28.

———. 1994. "Serious *Physalia* (Portuguese Man o' War) Stings: Implications for Scuba Divers." *Journal of Wilderness Medicine* 5 (1): 71–76.

———. 1986. "Venomous Stingray Injuries." *Cutis* 38 (August): 112.

———. 1990. "Verapamil Potentiation of *Chironex* (Box Jellyfish) Antivenom." *Toxicon* 28 (2): 242–244.

"Burns." Revised in 1992. Poisindex® Toxicologic Managements 84 Database online. Section E, "Sunburn." Denver: Micromedex Inc., 1974–1995.

Calderon, Rebecca, and Eric W. Mood. 1982. "An Epidemiological Assessment of Water Quality and Swimmer's Ear." *Archives of Environmental Health* 37 (September–October): 300–305.

Cameron, J., and M. F. Capra. 1993. "The Basis of the Paradoxical Disturbance of Temperature Perception in Ciguatera Poisoning." *Clinical Toxicology* 31: 571–579.

Campbell, George Duncan, and Edwin D. Smith. 1993. "The Problem of Shark Attacks Upon Humans." (Editorial) *Journal of Wilderness Medicine* 4 (1): 5–10.

"Causes and Consequences of Injury in Hawaii." 1994. Pacific Basin Rehabilitation Research and Training Center and the Hawaii State Department of Health, Injury Prevention and Control Program.

Centers for Disease Control and Prevention, U.S. Public Health Service. 1991. "Fish Botulism—Hawaii, 1990." *Journal of the American Medical Association* 266 (3): 324.

Centers for Disease Control and Prevention, U.S. Public Health Service. 1996. "*Vibrio vulnificus* Infections Associated with Eating Raw Oysters, Los Angeles, 1996." *Morbidity and Mortality Weekly Report* 45 (July): 621–624.

Centers for Disease Control and Prevention, U.S. Public Health Service. 1991. "Diphtheria, Tetanus, and Pertussis: Recommendations for the Vaccine Use and Other Preventive Measures." *Morbidity and Mortality Weekly Report* 40 (August): 21–22.

———. 1993. "Imported Cholera Associated with a Newly Described Toxigenic Vibrio Cholerae 0139 Strain—California, 1993." *Morbidity and Mortality Weekly Report* 42 (9 July): 213–219.

———. 1995. "Outbreak of Gastrointestinal Illness Associated with Consumption of Seaweed—Hawaii, 1994." *Morbidity and Mortality Weekly Report* 44 (6 October): 724–727.

Chang, Laurette A., and Clarence E. McDanal, Jr. 1980. "Boardsurfing and Bodysurfing Injuries Requiring Hospitalization in Honolulu." *Hawaii Medical Journal* 39 (May): 117.

Cheng, Charles L. Y., et al. 1992. "Bodysurfing Accidents Resulting in Cervical Spinal Injuries." *Spine* 17 (3): 257–260.

Ching, Patrick. 1994. *The Hawaiian Monk Seal*. Honolulu: University of Hawai'i Press.

Chisholm, C. D., et al. 1989. "Plantar Puncture Wounds: Controversies and Treatment Recommendations." *Annals of Emergency Medicine* 18 (December): 1,352.

Chrintz, H., et al. 1989. "Need for Surgical Wound Dressing." *British Journal of Surgery* 76 (February): 204–205.

Chu, George W. T. C., and Charles E. Cutress. 1955. "Dermatitis Due to Contact with the Hydroid, *Syncoryne mirabilis* (Agassiz, 1862)." *Hawaii Medical Journal* 14 (May–June): 403–405.

Chu, I., et al. 1982. "Toxicity of Trihalomethanes: The Acute and Subacute Toxicity of Chloroform, Bromodichloromethane, Chlorodibromomethane and Bromoform in Rats." *Journal of Environmental Science Health* 17 (3): 205–224.

"Ciguatera Fish Poisoning." 1992. Hawaii Department of Health, Health Promotion and Education Branch, Epidemiology Branch.

"Ciguatera Fish Poisoning." Revised in February 1994. Poisindex® Toxicologic Managements 84. Database online. Denver: Micromedex Inc., 1974–1995.

Clark, A. M., and J. K. Sims. 1989. *Dangerous Marine Organisms of Hawaii*. Revised edition. Sea Grant Advisory Report, UNIHI-SEAGRANT-AR-78-01.

Clench, William J., and Yoshio Kondo. 1943. "The Poison Cone Shell." *American Journal of Tropical Medicine* 23 (1): 105–121.

"Coelenterates." Revised in January 1994. Poisindex® Toxicologic Managements 84. Database online. Denver: Micromedex Inc., 1974–1995.

"Conotoxins." Revised in February 1992. Poisindex® Toxicologic Managements 84. Database online. Denver: Micromedex Inc., 1974–1995.

Cooper, Maxwell A. 1981. "Treatment of Coral Cuts in Hawaii." *Hawaii Medical Journal* 40 (March): 73–74.

"Coral." Revised in October 1992. Poisindex® Toxicologic Managements 84. Database online. Denver: Micromedex Inc., 1974–1995.

Coughenower, D. Douglas. 1994. "Get a Grip on Ocean Motion" (brochure). University of Alaska Sea Grant Marine Advisory Program. Homer, Alaska.

Coventry, D. M., et al. 1989. "Alkalinisation of Bupivacaine for Sciatic Nerve Blockade." *Anesthesia* 44 (June): 467.

Cummings, P. 1995. "Antibiotics to Prevent Infection of Simple Wounds: A Meta-Analysis of Randomized Studies." *American Journal of Emergency Medicine* 13 (July): 396.

Cummings, P., et al. 1994. "Antibiotics to Prevent Infection in Patients with Dog Bites: A Meta-Analysis of Randomized Trails." *Annals of Emergency Medicine* 23 (March): 535.

Currie, Bart J., and Yvonne K. Wood. 1995. "Identification of *Chironex fleckeri* Envenomation by Nematocyst Recovery from Skin." *Medical Journal of Australia* 162 (1 May): 478–480.

Dal-Re, R., et al. 1995. "Does Tetanus Immune Globulin Interfere with the Immune Response to Simultaneous Administration of Tetanus-Diphtheria Vaccine? A

Comparative Clinical Trial in Adults." *Journal of Clinical Pharmacology* 35 (April): 420.

Damron, Roy. N.d. "Scuba Fatalities—State of Hawaii, Through 1992." Unpublished.

Deardorff, Thomas L., et al. 1991. "Human Anisakiasis Transmitted by Marine Food Products." *Hawaii Medical Journal* 50 (January): 9–16.

———. 1986. "Invasive Anisakiasis: A Case Report from Hawaii." *Gasteroenterology* 90: 1,047–1,050.

"Decompression Sickness." Revised in March 1993. Poisindex® Toxicologic Managements 84. Database online. Denver: Micromedex Inc., 1974–1995.

Denoncourt, Paul M. 1991. "Seal Finger." *Orthopedics* 14 (June): 709–710.

De Sylva, Donald. 1963. "Systematics and Life History of the Great Barracuda." University of Miami Press, Institute of Marine Science. Ph.D. dissertation.

"Detergents—Cationic." Revised in April 1992. Poisindex® Toxicologic Managements 84. Database online. Denver: Micromedex Inc., 1974–1995.

Devaney, Dennis M., and Lucius G. Eldredge, eds. 1977. *Reef and Shore Fauna of Hawaii.* Section 1, *Protozoa through Ctenophora.* Honolulu: Bishop Museum Press.

———. *Reef and Shore Fauna of Hawaii.* 1987. Section 2, *Platyhelminthes through Phoronida,* and Section 3, *Sipuncula through Annelida.* Honolulu: Bishop Museum Press.

Davis, R. H., et al. 1989. "Oral and Topical Activity of Aloe Vera." *Journal of the Podiatric Medical Association* 79 (November): 559.

Dierauf, L. A., ed. 1990. *CRC Handbook of Marine Mammal Medicine: Health, Disease and Rehabilitation.* Boca Raton, Fla.: CRC Press.

Dire, D. J., et al. 1990. "A Comparison of Wound Irrigation Solutions Used in the Emergency Department." *Annals of Emergency Medicine* 19 (June): 704.

———. 1994. "A Prospective Evaluation of Risk Factors for Infections from Dog-Bite Wounds." *Academy of Emergency Medicine* 1 (May–June): 258.

———. 1995. "Prospective Evaluation of Topical Antibiotics for Preventing Infections in Uncomplicated Soft-Tissue Wounds Repaired in the ED." *Academic Emergency Medicine* 2 (January): 4.

Doury, P. 1990. "Is There a Role for Parasites in the Etiology of Inflammatory Rheumatism?" *Bulletin of Academic National Medicine* 174 (June–July): 751–754.

Doweck, Ilana, et al. 1994. "Effect of Cinnarizine in the Prevention of Seasickness." *Aviation, Space and Environmental Medicine* 65 (July): 606–609.

Dovenbarger, Joel. 1995. "The Study of Fatalities in Recreational Scuba Diving." *Alert Diver* (September–October): 28–31.

Dovenbarger, Joel, and Donna Uguccioni. 1995. "1994 Injuries in Recreational Diving: A Preview of DAN's Upcoming Report on Diving Accidents and Fatalities." *Alert Diver* (November–December): 22–25.

Eadie, P. A., et al. 1990."Seal Finger in a Wildlife Ranger." *Irish Medical Journal* 83 (September): 117–118.

Eastaugh, Janet A. 1996. "Delayed Use of Intravenous Mannitol in Ciguatera (Fish Poisoning)." Letter to the editor. *Annals of Emergency Medicine* 28 (July): 105.

Edelstein, Howard. 1994. "*Mycobacterium marinum* Skin Infections. Report of 31 Cases and Review of the Literature." *Archives of Internal Medicine* 154 (June): 1,359–1,364.

Edmonds, Carl. 1995. *Dangerous Marine Creatures*. Flagstaff, Ariz.: Best Publishing.

Erickson, Tim, et al. 1992. "The Emergency Management of Moray Eel Bites." *Annals of Emergency Medicine* 21 (February): 212–216.

Exton, David R., et al. 1989. "Cold Packs: Effective Topical Analgesia in the Treatment of Painful Stings by *Physalia* and Other Jellyfish." *Medical Journal of Australia* 151 (4–18 December): 625–626.

Fariss, B. L., et al. 1987. "Anesthetic Properties and Toxicity of Bupivacaine and Lidocaine for Infiltration Anesthesia." *Journal of Emergency Medicine* 5: 275.

Fenner, Peter. J. 1995. "Stingray Envenomation: A Suggested New Treatment." Letter in *Medical Journal of Australia* 163 (4–18 December): 655.

Fenner, Peter J., et al. 1989. "Fatal and Non-fatal Stingray Envenomation." *Medical Journal of Australia* 151 (4–18 December): 621–624.

———. 1993. "First Aid Treatment of Jellyfish Stings in Australia." *Medical Journal of Australia* 158 (April): 498–501.

Fielding, Ann. 1987. *An Underwater Guide to Hawaii*. Honolulu: University of Hawai'i Press.

———. 1985. *Hawaiian Reefs and Tidepools*. Honolulu: Oriental Publishing.

"Fish Poisoning in Hawaii." 1991. Brochure. Hawaii Department of Health, Epidemiology Branch.

"Fish Stings." Revised in January 1994. Poisindex® Toxicologic Managements 84. Database online. Denver: Micromedex Inc., 1974–1995.

"Fisherman Slashed by Barracuda." 1965. *Honolulu Star-Bulletin and Advertiser,* 12 September, p. A-1.

Fortenberry, J. Eric, et al. 1995. "Use of Epinephrine for Anaphylaxis by Emergency Medical Technicians in a Wilderness Setting." *Annals of Emergency Medicine* 25 (June): 785–787.

Francis, T. J. R., and A. J. Dutka. 1989. "Methyl Prednisolone in the Treatment of Acute Spinal Cord Decompression Sickness." *Undersea Biomedical Research* 16 (2): 165–174.

Garcia, Patricia J., et al. 1994. "Fulminant Hepatic Failure from a Sea Anemone Sting." *Annals of Internal Medicine* 120: 665–666.

"Gastroenteritis—Cholera." Revised in 1994. Poisindex® Toxicologic Managements 84. Database online. Denver: Micromedex Inc., 1974–1995.

Girard, Susan M., et al. 1979. "*Clostridium perfingens* cultured from a Hawaiian Sardine, *Sardinella marquesesis.*" *Hawaii Medical Journal* 38 (November): 327–329.

Glasser, D. B., et al. 1992. "Ocular Jellyfish Stings." *Ophthalmology* 99: 1,414–1,418.

Goldstein, Norman. 1993. "Skin Cancers in Hawaii (1993)." *Hawaii Medical Journal* 52 (May): 126–128.

Gollop, James H., and Eugene W. Pon. 1992. "Ciguatera: A Review." *Hawaii Medical Journal* 51 (April): 91–99.

Gordon, Carlos R., et al. 1994. "Seasickness Susceptibility, Personality Factors and Salivation." *Aviation, Space and Environmental Medicine* 65 (July): 610–614.

Graf, William D., et al. 1995. "Predicting Outcome in Pediatric Submersion Victims." *Annals of Emergency Medicine* 26 (September): 312–319.

Grauer, Colonel Franklin H. 1959. "Dermatitis Escharotica Caused by a Marine Alga." *Hawaii Medical Journal* 19 (September–October): 32–34.

Green, Steven M., et al. 1993. "Incidence and Severity of Middle Ear Barotrauma in Recreational Scuba Diving." *Journal of Wilderness Medicine* 4 (3): 270–280.

Grontved, Aksel, and Erwin Hentzer. 1986. "Vertigo-Reducing Effect of Ginger Root: A Controlled Clinical Study." *ORL* 48 (5): 282–286.

Grontved, Aksel, et al. 1988. "Ginger Root Against Seasickness—A Controlled Trial on the Open Sea." *Acta Oto-Laryngologica* 105 (January–February): 45–49.

Halstead, Bruce W. 1987. "Coelenterate (Cnidarian) Stings and Wounds." *Clinics in Dermatology* 5 (July–September): 8–13.

———. 1992. *Dangerous Aquatic Animals of the World: A Color Atlas (with Prevention, First Aid and Emergency Treatment Procedures).* Princeton, N.J.: Darwin Press.

———. 1988. *Poisonous and Venomous Marine Animals of the World.* 2nd rev. ed. Princeton, N.J.: Darwin Press.

Halstead, Bruce W., et al. 1990. *A Color Atlas of Dangerous Marine Animals.* Boca Raton, Fla.: CRC Press.

Harth, Manfred, et al. 1994. "Septic Arthritis Due to *Mycobacterium marinum*." *Journal of Rheumatology* 21 (May): 957–960.

Hartung, G., and Deborah A. Goebert. N.d. "Ocean Sports and Recreational Activities." Hawaii State Department of Health, Injury Prevention and Control Program. Unpublished.

Hartung, G. Harley, et al. 1990. "Epidemiology of Ocean Sports–Related Injuries in Hawaii: Akahele O Ke Kai." *Hawaii Medical Journal* 49 (February): 52–56.

Hartwick, Robert, et al. 1980. "Disarming the Box Jellyfish." *Medical Journal of Australia* 1 (12 January): 15–20.

Hashimoto, Yoshiro, et al. 1969. "Aleuterin: A Toxin of Filefish, *Alutera scripta*, Probably Originating from a Zoantharian, *Palythoa tuberculosa*." *Bulletin of the Japanese Society of Scientific Fisheries* 35 (11): 1,086–1,093.

Helfrich, Philip. 1963. "Fish Poisoning in Hawaii." *Hawaii Medical Journal* 22 (May–June): 361–372.

Helfrich, Philip, and Albert H. Banner. 1960. "Hallucinatory Mullet Poisoning." *Tropical Medicine and Hygiene* 63 (April): 86–89.

Hinegardner, R. T. 1958. "The Venom Apparatus of the Cone Shell." *Hawaii Medical Journal* 17 (July–August): 533–536.

Hiramoto, Joy T., and Jinichi Tokeshi. 1991. "Anisakiasis in Hawaii: A Radiological Diagnosis." *Hawaii Medical Journal* 50 (June): 202–203.

Hlady, Gary W., et al. 1993. "*Vibrio vulnificus* from Raw Oysters: Leading Cause of Reported Deaths from Food-borne Illness in Florida." *Journal of the Florida Medical Association* 80 (August): 536–538.

Hobson, Edmund, and E. H. Chave. 1990. *Hawaiian Reef Animals.* Rev. ed. Honolulu:

University of Hawai'i Press.

Hong, Jennifer. 1995. "Pair Predicted Flotilla's Return." *Honolulu Advertiser,* 24 May, p. A-1.

Hoover, John P. 1993. *Hawaii's Fishes: A Guide for Snorkelers, Divers and Aquarists.* Honolulu: Mutual Publishing.

Howell, John M., et al. 1993. "The Effect of Scrubbing and Irrigation with Normal Saline, Povidone Iodine and Cefazolin on Wound Bacterial Counts in a Guinea Pig Model." *American Journal of Emergency Medicine* 11 (March): 134.

————. 1993. "The Effect of Scrubbing and Irrigation on Staphylococcal and Streptococcal Counts in Contaminated Lacerations." *Antimicrobial Agents and Chemotherapy* 37 (December): 2,754–2,755.

Hughes, G. S., et al. 1992. "Synergistic Effects of Oral Nonsteroidal Drugs and Topical Corticosteroids in the Therapy of Sunburn in Humans." *Dermatology* 184 (1): 54.

Ikeda, Tomosuma. 1989. "Supraventricular Bigeminy Following a Stingray Envenomation: A Case Report." *Hawaii Medical Journal* 45 (5): 162–164.

Iredell, Jon, et al. 1992. "*Mycobacterium marinum* Infection: Epidemiology and Presentation in Queensland 1971–1990." *Medical Journal of Australia* 157 (2 November): 596–598.

Isakov, Alexander P., et al. 1996. "Acute Carpal Tunnel Syndrome in a Diver: Evidence of Peripheral Nervous System Involvement in Decompression Illness." *Annals of Emergency Medicine* 28 (July): 90–93.

Izumi, Allan K., and Richard E. Moore. 1987. "Seaweed *(Lyngbya majuscula)* Dermatitis." *Clinics in Dermatology* 5 (July–September): 92–100.

Johnson, Scott. N.d. *Living Seashells.* Honolulu: Oriental Publishing.

Johnston, Jan M., and Allan K. Izumi. 1987. "Cutaneous *Mycobacterium marinum* Infection (Swimming Pool Granuloma)." *Clinics in Dermatology* 5 (July–September): 68–75.

Johnston, Jeffery M., et al. 1985. "*Vibrio vulnificus*—Man and the Sea." *Journal of the American Medical Association* 253 (17 May): 2,850–2,853.

Kaplan, Lee A. 1992. "Suntan, Sunburn, and Sun Protection." *Journal of Wilderness Medicine* 3 (2): 173–196.

Kasdan, Morton L., et al. 1987. "Lionfish Envenomation of the Hand." *Plastic and Reconstructive Surgery* 80 (October): 613–614.

Kaufman, Michele B. 1992. "Portuguese Man-of-War Envenomation." *Pediatric Emergency Care* 8 (February): 27–28.

Kay, E. Allison. 1979. *Hawaiian Marine Shells. Reef and Shore Fauna of Hawaii,* Section 4, *Mollusca.* Honolulu: Bishop Museum Press.

Kimura, Shoji, and Yoshiro Hashimoto. 1973. "Purification of the Toxin of a Zoanthid." *Publications of the Seto Marine Biological Laboratory* 20 (December): 713–718.

Kimura, Shoji, and Kiyoshi Yamazato. 1972. "Toxicity of the Zoanthid *Palythoa tuberculosa.*" *Toxicon* 10: 611–617.

King, D. Friday, and Victor D. Newcomer. 1982 "Sunburn, Aging and the Carcinogenic Effects of Sunlight on the Skin." *Hawaii Medical Journal* 41 (November): 400–402.

Kizer, Kenneth W., et al. 1985. "Scorpaenidae Envenomation: A Five-Year Poison Center Experience." *Journal of the American Medical Association* 253 (8 February): 807–810.

Klontz, Karl C., et al. 1993. "*Vibrio* Wound Infections in Humans Following Shark Attack." *Journal of Wilderness Medicine* 4: 68–72.

Kodama, Arthur M., et al. 1989. "Clinical and Laboratory Findings Implicating Palytoxin as Cause of Ciguatera Poisoning Due to *Decapterus macrosoma* (Mackerel)." *Toxicon* 27 (9): 1,051–1,053.

Koerner, Michael. 1993. "Guillain-Barré Syndrome after Sea Urchin Envenomation." *Journal of Wilderness Medicine* 4 (4): 463–464.

Kohn, A. J. 1958. "Cone Shell Stings, Recent Cases of Human Injury Due to Venomous Marine Snails of the Genus *Conus*." *Hawaii Medical Journal* 17 (July–August): 528–532.

Kohn, A. J. 1963. "Venomous Marine Snails of the Genus *Conus*," in *Venomous and Poisonous Animals and Noxious Plants of the Pacific Area*. Oxford, U.K.: Percamon Press.

Kohn, A. J., et al. 1960. "Preliminary Studies on the Venom of the Marine Snail *Conus*." *Annals of the New York Academy of Sciences* 90 (17 November): 706–725.

Kol, Shahar, et al. 1993. "Pulmonary Barotrauma After a Free Dive—A Possible Mechanism." *Aviation, Space and Environmental Medicine* (March): 236–237.

Kontoyiannis, Dimitrios P., et al. 1995. "Primary Septicemia Caused by *Vibrio cholerae* Non-01 Acquired on Cape Cod, Massachusetts." *Clinical Infectious Diseases* 21 (November): 1,330–1,333.

Koriwchak, Michael J., and Jay A. Werkhaven. 1994. "Middle Ear Barotrauma in Scuba Divers." *Journal of Wilderness Medicine* 5 (4): 389–398.

Krause, M. 1992. "Infectious Sinusitis and Otitis." *Ther-Umsch* 49 (April): 216–221.

Lane, Charles E. 1960. "The Portuguese Man-of-War." *Scientific American* 202: 158–168.

Lange, W. Robert. 1990. "Puffer Fish Poisoning." *American Family Physician* 42 (October): 1,029–1,033.

Lawther, Anthony, and Michael J. Griffin. 1988. "A Survey of the Occurrence of Motion Sickness Amongst Passengers at Sea." *Aviation, Space and Environmental Medicine* 59 (May): 399–406.

LeCurieux F., et al. 1995. "Use of the SOS Chromotest, the Ames-Fluctuation Test and the Newt Micronucleus Test to Study the Genotoxicity of Four Trihalomethanes." *Mutagenesis* 10 (July): 333–341.

Levy, Shlomo, et al. 1970. "Report of Stingings by the Sea Anemone *Triactis producta* Klunzinger from the Red Sea." *Clinical Toxicology* 3 (December): 637–643.

Limpert, George H., and James E. Peacock. 1988. "Soft Tissue Infections Due to Noncholera Vibrios." *American Family Practice* 37 (February): 193–198.

Limpus, Colin. 1987. "Sea Turtles," in *Toxic Plants and Animals: A Guide for Australia*, ed. Jeanette Covacevich et al., pp. 189–193.

Lineaweaver, W., et al. 1985. "Topical Antimicrobial Toxicity." *Archives of Surgery* 120 (March): 267.

Lucot, James B., and George H. Crampton. 1991. "Pharmacological and

Neurophysiological Aspects of Space/Motion Sickness." NASA Contractor Report.

Lumley, John, et al. 1988. "Fatal Envenomation by *Chironex fleckeri*, the North Australian Box Jellyfish: The Continuing Search for Lethal Mechanisms." *Medical Journal of Australia* 148 (16 May): 527–534.

McCabe, M. J., et al. 1978. "A Fatal Brain Injury Caused by a Needlefish." *Neuroradiology* 15: 137–139.

McIntosh, J. M., et al. 1993. "Presence of Serotonin in the Venom of *Conus Imperialis*." *Toxicon* 31 (December): 1,561–1,566.

McKeown, Sean. 1978. *Hawaiian Reptiles and Amphibians*. Honolulu: Oriental Publishing.

Magruder, William H., and Jeffery W. Hunt. 1979. *Seaweeds of Hawaii*. Honolulu: Oriental Publishing.

Maimaris, C., et al. 1988. "Dog-Bite Lacerations: A Controlled Trial of Primary Wound Closure." *Archives of Emergency Medicine* 5 (September): 156.

Mandojana, Ricardo M., and Joel K. Sims. 1987. "Miscellaneous Dermatoses Associated with the Aquatic Environment." *Clinics in Dermatology* 5 (July–September): 134–145.

Maretic, Zvonimir, and Findlay E. Russell. 1983. "Sting by the Sea Anemone *Anemonia sulcata* in the Adriatic Sea." *American Journal of Tropical Medicine* 32 (4): 891–896.

Markham, Richard B., and B. Frank Polk. 1979. "Seal Finger." *Reviews of Infectious Diseases* 1 (May–June): 567–569.

Marshall, Katherine. 1995. "Seaweed Toxicity." *Communicable Disease Report*, Hawaii Department of Health, Communicable Disease Division. March–April: 1–2.

Mass, Daniel P., et al. 1981. "Seal Finger." *Journal of Hand Surgery* 6 (November): 610–612.

Mathelier-Fusade P., and F. Leyandier. 1993. "Acquired Cold Urticaria after Jellyfish Sting." *Contact Dermatitis* 29: 273.

Matsumoto, G. I. 1995. "Observation on the Anatomy and Behavior of the Cubozoan *Carybdea rastonii* Haacke." *Mar. Fresh. Behav. Physiol.* 26: 139–148.

Meadors, Michael C., and George A. Pankey. 1990. "*Vibrio vulnificus* Wound Infection Treated Successfully with Oral Ciprofloxacin." *Journal of Infection* 20 (January): 88.

Meier, Jurg, and Julian White, ed. 1995. *Clinical Toxicology of Animal Venoms*. Boca Raton, Fla.: CRC Press.

Melton, Robert J., et al. 1984. "Fatal Sardine Poisoning." *Hawaii Medical Journal* 43 (April): 114–124.

Mettler, R., and K. Biener. 1991. "Athletic Injuries in Windsurfing." *Schweiz-Z-Sportmed* 39 (December): 161–166.

Miller, Bruce J., and Scott P. Bogle. 1993. "Ozone Depletion: Causes, Potential Effects and Remedies." *Hawaii Medical Journal* 52 (May): 118–122.

Millington, J. Thomas, and Peggy Wilhelm. 1993. "Marine Microbiology of Rocas Alijos." *Journal of Wilderness Medicine* 4 (4): 384–390.

Mintz, Eric D., et al. 1994. "A Rapid Public Health Response to a Cryptic Outbreak of Cholera in Hawaii." *American Journal of Public Health* 84 (December): 1988–1991.

Moon, Richard E., et al. 1995. "The Physiology of Decompression Sickness." *Scientific American* 273 (August): 70–77.

Morris, J. Glenn, et al. 1985. "Cholera and Other Vibrioses in the United States." *New England Journal of Medicine* 312 (7 February): 343–350.

Morrow, J. D., et al. 1991. "Evidence That Histamine is the Causative Toxin in Scombroid Fish Poisoning." *New England Journal of Medicine* 324 (14 March): 716.

Moscati, R., et al. 1990. "Comparison of Cimetidine and Diphenhydramine in the Treatment of Acute Urticaria." *Annals of Emergency Medicine* 19 (January): 12.

Mowrey, Daniel B., and Dennis E. Clayson. 1982. "Motion Sickness, Ginger, and Psychophysics." *Lancet* 1 (20 March): 655–657.

Myers, Robert F. 1989. *Micronesian Reef Fishes: A Practical Guide to the Identification of the Coral Reef Fishes of the Tropical Central and Western Pacific*. Guam: Coral Graphics.

Nakata, M. 1994. "Scombroid Fish Poisoning: It's Not a Food Allergy." *Communicable Disease Report*, Hawaii Department of Health, Communicable Disease Division. September–October, p. 2.

"Near Drowning." Revised in September 1994. Emergindex® Toxicologic Managements 84. Database online. Denver: Micromedex Inc., 1974–1995.

Newmeyer, William L. 1988. "Management of Sea Urchin Spines in the Hand." *Journal of Hand Surgery* 13A (May): 455–457.

Nichter, M. A. 1989. "Childhood Near-Drowning: Is Cardiopulmonary Resuscitation Always Indicated?" *Critical Care Medicine* 17 (October): 993.

"1993 Report on Diving Accidents and Fatalities." 1995. Divers Alert Network, Duke University Medical Center, Durham, N.C.

Odom, Charles B. 1984. "More on Salt Water Drowning." *Hawaii Medical Journal* 43 (July): 218.

Odom, Charles B., and Edward A. Fischermann. 1971. "Crown-of-Thorns Starfish Wounds—Some Observations on Injury Sites." *Hawaii Medical Journal* 31 (March–April): 99–100.

O'Donnell, B. F., and C. Y. Tan. 1993. "Persistent Contact Dermatitis from Jellyfish Sting." *Contact Dermatitis* 28: 112.

Ohtaki, Norito, et al. 1990. "Cutaneous Reactions Caused by Experimental Exposure to Jellyfish, *Carybdea rastonii*." *Journal of Dermatology* 17: 108–114.

"Ondansetron." Revised in 1995. Drug Evaluation Monograph. Emergindex® Toxicologic Managements 84. Database online. Denver: Micromedex Inc., 1974–1995.

"Otitis Externa." Revised in March 1992. Emergindex®Toxicologic Managements 84. Database online. Denver: Micromedex Inc., 1974–1995.

Pacelli, Lauren C. 1990. "Body Surfing: Fun and Free, but Potentially Dangerous." *The Physician and Sports Medicine* 18 (September): 145–149.

Palafox, Neal A., et al. In press. "Intravenous 20% Mannitol versus Intravenous 5% Dextrose for the Treatment of Acute Ciguatera: A Randomized, Placebo Controlled, Double Masked Clinical Trial."

———. 1988. "Successful Treatment of Ciguatera Fish Poisoning with Intravenous

Mannitol." *Journal of the American Medical Association* 269: 2,740–2,742.

"Palytoxin." Revised in March 1994. Poisindex® Toxicologic Managements 84. Database online. Denver: Micromedex Inc., 1974–1995.

Parell, G. Joseph, and Gary D. Becker. 1993. "Inner Ear Barotrauma in Scuba Divers." *Archives of Otolaryngology, Head, and Neck Surgery* 119:455–457.

Parent, Leslie J., et al. 1995. "Disseminated *Mycobacterium marinum* Infection and Bacteremia in a Child with Severe Combined Immunodeficiency." *Clinical Infectious Diseases* 21 (November): 1,325–1,327.

Patel, M. R., and Scott Wells. 1993. "Lionfish Envenomation of the Hand." *Journal of Hand Surgery* 18A (May): 523–525.

Paul, Devchand, and Peter Gulick. 1993. "*Mycobacterium marinum* Skin Infections: Two Case Reports." *Journal of Family Practice* 36 (3): 336–338.

Pearse, Vicki, John Pearse, Mildred Buchsbaum, and Ralph Buchsbaum. 1987. *Living Invertebrates*. Palo Alto, Calif.: Blackwell Scientific Publications and Pacific Grove, Calif.: Boxwood Press.

Pien, Francis, et al. 1977. "*Vibrio alginolyticus* Infections in Hawaii." *Journal of Clinical Microbiology* 5 (June): 670–672.

Pingree, B. J. W. 1994. "INM Investigations into Drugs for Seasickness Prophylaxis." *Journal of the Royal Naval Medical Services* 80 (Summer): 76–80.

Pingree, B. J. W., and R. J. Pethybridge. 1994. "Motion Commotion—A Seasickness Update." *Journal of the Royal Naval Medical Services* 75 (Summer): 75–84.

Pingree, B. J. W., 1989. "A Comparison of the Efficacy of Cinnarizine with Scopolomine in the Treatment of Seasickness." *Aviation, Space and Environmental Medicine* 65 (July): 597–605.

Pollack, C., et al. 1995. "Outpatient Management of Acute Urticaria: The role of Prednisone." *Annals of Emergency Medicine* 26 (November): 547.

Pongprayoon, U., et al. 1991. "Neutralization of Toxic Effects of Different Crude Jellyfish Venoms by an Extract of *Ipomoea pes-carprae*." *Journal of Ethnopharmacology* 35: 65–69.

Pratt, Franklin D., and Bruce E. Haynes. 1986. "Incidence of Secondary Drowning after Saltwater Submersion." *Annals of Emergency Medicine* 15 (September): 1,084–1,087.

"Preventing Childhood Injury: Hawaii's Strategic Plan 1995–2000." 1995. Injury Prevention and Control Program, Hawaii Department of Health.

"Preventing Serious Raw Oyster–Associated *Vibrio vulnificus* Infections." 1995. *FDA Medical Bulletin,* August, p. 4.

"Protocol for the Investigation of Fish Poisoning." 1995. Hawaii Department of Health, Epidemiology Branch, August (unpublished).

"Putting Sunscreens to the Test." 1995. *Consumer Reports,* May, pp. 332–339.

Quan, Linda. 1993. "Drowning Issues in Resuscitation." *Annals of Emergency Medicine* 22, pt. 2 (February): 100–103.

Quan, Linda, and D. Kinder. 1992. "Pediatric Submersions: Prehospital Predictors of Outcome." *Pediatrics* 90: 909–913.

Quan, Linda, et al. 1990. "Outcome and Predictors of Outcome in Pediatric Submersion

Victims Receiving Prehospital Care in King County, Washington." *Pediatrics* 86 (October): 586.

Randall, John E. 1985. *Guide to Hawaiian Reef Fishes.* Kaneohe, Hawai'i: Treasures of Nature.

Randall, John E., et al. 1990. *Fishes of the Great Barrier Reef and Coral Sea.* Honolulu: University of Hawai'i Press.

———. 1981. "Occurrence of a Crinotoxin and Hemagglutinin in the Skin Mucus of the Moray Eel, *Lycodontis nudivomer.*" *Marine Biology* 62: 179–184.

Rifkin, Jacqueline, et al. 1993. "First Aid of the Sting from the Hydroid *Lytocarpus philippinus:* The Structure of, and In Vitro Discharge Experiments with, Its Nematocysts." *Journal of Wilderness Medicine* 4 (3): 252–260.

———. 1993. Reply to "Disarming the Box Jellyfish." *Medical Journal of Australia* 158 (May): 647.

Rippey, Scott R. 1994. "Infectious Diseases Associated with Molluscan Shellfish Consumption." *Clinical Microbiology Review* 7 (October): 419–425.

Rheinstein, Peter H., and Karl C. Klontz. 1993. "Shellfish-Borne Illness." *American Family Physician* 47 (June): 1,837–1,840.

Rodahl, Kare. 1943. "Notes on the Prevention and Treatment of Spekk Finger." *Polar Record* 4 (January): 17–19.

Rodrigue-Bigas, M., et al. 1988. "Comparative Evaluation of Aloe Vera in the Management of Burn Wounds in Guinea Pigs." *Plastic Reconstructive Surgery* 81 (March): 386.

Rolnick, A., and W. Bles. 1989. "Performance and Well-Being Under Tilting Conditions: The Effects of Visual Reference and Artificial Horizon." *Aviation, Space and Environmental Medicine* 60 (August): 779–785.

Rosen, R. A. 1985. "The Use of Antibiotics in the Initial Management of Recent Dog-Bite Wounds." *American Journal of Emergency Medicine* 3 (January): 19.

Russell, F. E. 1984. "Marine Toxins and Venomous and Poisonous Marine Plants and Animals (Invertebrates)." *Advances in Marine Biology* 21: 59–192.

———. 1953. "Stingray Injuries: A Review and Discussion of Their Treatment." *American Journal of the Medical Sciences* 226: 611–622.

Sanford, Jay P., et al. 1996. "Guide to Antimicrobial Therapy." Dallas: Antimicrobial Therapy, Inc.

Sargent, Ed. 1980. "Tetracycline for Seal Finger." Letter in *Journal of the American Medical Association* 244 (1 August): 437.

Schmedtje, John F., et al. 1988. "Effects of Scopolomine and Dextroamphetamine on Human Performance." *Aviation, Space and Environmental Medicine* 59 (May): 407–410.

Schwartz, Karen. 1996. "Dental Work and Diving." *Alert Diver* (July–August): 31.

"Scombroid Fish Poisoning." Revised in February 1994. Poisindex® Toxicologic Managements 84. Database online. Denver: Micromedex Inc., 1974–1995.

Scott, Susan. 1993. *Exploring Hanauma Bay.* Honolulu: University of Hawai'i Press.

———. 1988. *Oceanwatcher: An Above-Water Guide to Hawaii's Marine Animals.* Honolulu: Green Turtle Press.

———. 1991. *Plants and Animals of Hawaii.* Honolulu: Bess Press.

"Sea Snakes." Revised in November 1991. Poisindex® Toxicologic Managements 84. Database online. Denver: Micromedex Inc., 1974–1995.

"Seaweed Burn." 1959. *Hawaii Medical Journal* 18 (July–August): 603.

Sepulveda, J., et al. 1992. "Cholera in the Americas." *Infection* 20 (5): 243–248.

Serdula, Mary, et al. 1982. "Seaweed Itch on Windward Oahu." *Hawaii Medical Journal* 41 (July): 200–201.

Shirahama, M., et al. 1992. "Intestinal Anisakiasis: US is Diagnostic." *Radiology* 185: 789–793.

Shirai, J. L., et al. 1991. *Seafood Poisoning: Ciguatera.* Gardena, Calif.: Yosh Hokama Family Trust.

Shiroma, N., et al. 1994. "Haemodynamic and Haematologic Effects of *Acanthaster planci* Venom in Dogs." *Toxicon* 32 (October): 1,217–1,225.

———. 1990. "Liver Damage by the Crown-of-Thorns Starfish *(Acanthaster planci)* Lethal Factor." *Toxicon* 28 (5): 469–475.

Silas, E. G., and A. Bastian Fernando. 1984. "Turtle Poisoning." *Central Marine Fisheries Research Institute,* P.B. no. 1912, Cochin 682018, India. February, pp. 62–75.

Sims, J. K. 1987. "A Theoretical Discourse on the Pharmacology of Toxic Marine Ingestions." *Annals of Emergency Medicine* 16 (September): 1,006–1,015.

Sims, J. K., and Michel Y. Irei. 1979. "Human Hawaiian Marine Sponge Poisoning." *Hawaii Medical Journal* 38 (September): 263–270.

Sims, J. K., and Douglas Clark Ostman. 1986. "Pufferfish Poisoning: Emergency Diagnosis and Management of Mild Human Tetrodotoxication." *Annals of Emergency Medicine* 15 (September): 1,094–1,098.

Sims, J. K., and Richard D. Zandee Van Rilland. 1981. "Escharotic Stomatitis Caused by the 'Stinging Seaweed' *Microcoleus lyngbyaceus* (formerly *Lynbya majuscula*). *Hawaii Medical Journal* 40 (September): 243–248.

Sims, J. K., et al. 1993. "*Vibrio* in Stinging Seaweed: Potential Infection." *Hawaii Medical Journal* 52 (October): 274–275.

Skinner, John S. 1957. "Seal Finger." *A.M.A. Archives of Dermatology* 75 (April): 559–561.

Southcott, R. V., and J. R. Coulter. 1971. "The Effects of the Southern Australian Marine Stinging Sponges *Neofibularia mordens* and *Lissodendoryx* spp." *Medical Journal of Australia* 2: 895–901.

Stadtlander, Christian T. K.-H., and Sarabelle Madoff. 1994. "Characterization of Cytopathogenicity of Aquarium Seal Mycoplasmas and Seal Finger Mycoplasmas by Light and Scanning Electron Microscopy." *Zbl. Bakt.* 280: 458–467.

"Starfish Poisoning." Revised in February 1991. Poisindex® Toxicologic Managements 84. Database online. Denver: Micromedex Inc., 1974–1995.

Stein, Mark R., et al. 1989. "Fatal Portuguese Man-of-War *(Physalia physalis)* Envenomation." *Annals of Emergency Medicine* 18 (March): 131–134.

"Sting-Aid." Technical leaflet. 1993. Knight Industries, P.O. Box 50387, Pompano Beach, FL 33074.

Strauss, Michael B., and Rodger L. Dierker. 1987. "Otitis Externa Associated with Aquatic Activities (Swimmer's Ear)." *Clinics in Dermatology* 5 (July–September): 103–111.

Swift, A. E. B., and T. R. Swift. 1993. "Ciguatera." *Clinical Toxicology* 31: 1–29.

Tabrah, Frank L., et al. 1994. "Baromedicine Today—Rational Uses of Hyperbaric Oxygen Therapy." *Hawaii Medical Journal* 53 (April): 112–119.

Tacket, Carol O., et al. 1984. "Clinical Features and an Epidemiological Study of *Vibrio vulnificus* Infections." *Journal of Infectious Diseases* 149 (April): 558–561.

Taylor, Leighton. 1993. *Sharks of Hawaii: Their Biology and Cultural Significance.* Honolulu: University of Hawai'i Press.

TenBruggencate, Jan. 1996. "Swordfish Have a Weapon to Fight Back." *Honolulu Advertiser,* 17 March, p. A-1.

Tester, Albert L. 1969. "Cooperative Shark Research and Control Program, Final Report, 1967–1969." University of Hawai'i, 31 December.

"Tetanus Toxoid." Revised in October 1994. Emergindex® Toxicologic Managements 84. Database online. Denver: Micromedex Inc., 1974–1995.

"Tetrodotoxin." Revised in January 1992. Poisindex® Toxicologic Managements 84. Database online. Denver: Micromedex Inc., 1974–1995.

Thompson, Philip J., et al. 1993. "Seals, Seal Trainers, and Mycobacterial Infection." *American Review of Respiratory Disease* 147:164–167.

Tinker, Spencer Wilkie. 1982. *Fishes of Hawaii: A Handbook of the Marine Fishes of Hawaii and the Central Pacific Ocean.* Honolulu: Hawaiian Service.

Titcomb, Margaret, with M. K. Pukui. 1972. *Native Use of Fish in Hawaii.* Honolulu: University of Hawai'i Press.

"To Get a Fisherman Off the Hook." 1992. *Emergency Medicine,* 15 April, pp. 179–180.

Tomchik, Robert S., et al. 1993. "Clinical Perspectives on Seabather's Eruption, Also Known As Sea Lice." *Journal of the American Medical Association* 269 (7 April): 1,669–1,672.

Tong, David W. 1996. "Skin Hazards of the Marine Aquarium Industry." *International Journal of Dermatology* 35 (March): 153–158.

Tu, Anthony T. 1987. "Biotoxicology of Sea Snake Venom." *Annals of Emergency Medicine* 16 (September): 1,023–1,028.

Turner, Beryl, et al. 1980. "Disarming the Bluebottle: Treatment of *Physalia* Envenomation." *Medical Journal of Australia* 2: 394–395.

"Underwater Diving Accident Manual." 1994. Divers Alert Network, Duke University Medical Center, Durham, N.C.

Valentijn, K., et al. 1992. "Omega-Conotoxin- and Nifedipine-Insensitive Voltage-Operated Calcium Channels Mediate K (+)-Induced Release of Pro-Thyrotropin-Releasing Hormone-Connecting Peptides Ps4 and Ps5 from Perifused Rat Hypothalamic Slices." *Brain Research. Molecular Brain Research* 14 (July): 221–230.

Vann, R. D., et al. 1993. "Flying after Diving and Decompression Sickness." *Aviation, Space and Environmental Medicine* 64: 801–807.

Visuthikosol, V., et al. 1995. "Effect of Aloe Vera Gel to Healing of Burn Wounds in a

Clinical and Histological Study." *Journal of the Medical Association of Thailand* 78 (August): 403–409.

Walker, H. H., et al. 1962. "Some Characteristics of Swimming Pool Disease in Hawaii." *Hawaii Medical Journal* 21: 403–409.

Warpinski, James R., et al. 1993. "Fish Surface Mucin Hypersensitivity." *Journal of Wilderness Medicine* 4 (3): 261–269.

Warwick-Evans, L. A., et al. 1991. "A Double-Blind, Placebo-Controlled Evaluation of Acupressure in the Treatment of Motion Sickness." *Aviation, Space and Environmental Medicine* 62: 776–778.

Westerdahl, J., et al. 1994. "At What Age Do Sunburn Episodes Play a Crucial Role for the Development of Malignant Melanoma?" *European Journal of Cancer* 30A (11): 1,647–1,654.

Williamson, John, et al., eds. In press. *The Marine Stingers Reference Book.* Sydney, Australia: University of New South Wales Press.

Willingham, Barbara, and Joel Dovenbarger. 1996. "Understanding Diving: Vital Statistics." *Alert Diver* (July–August): 31.

Wong, Douglas E., et al. 1994. "Seabather's Eruption." *Journal of the Academy of Dermatology* 30 (March): 399–406.

Wong, Linda L., and J. Judson McNamara. 1984. "Salt Water Drowning." *Hawaii Medical Journal* 43 (June): 208–210.

Yaffee, Howard S. 1970. "Irritation from Red Sponge." *New England Journal of Medicine* 282(1): 51.

Yamada, Gregg M. 1993."A Rare Case of Cholera in Hawaii." *Hawaii Medical Journal* 52 (March): 62–64.

Yamamoto, Loren G., et al. 1992. "A One-Year Series of Pediatric ED Water-Related Injuries: The Hawaii EMS-C Project." *Pediatric Emergency Care* 8 (June): 129–133.

Yardon, Michael, et al. 1995. "Doctor, I'm Hooked." Letter in *Wilderness and Environmental Medicine* 6 (3): 348.

Zlotnick, Bradley A., et al. 1995. "Ciguatera Poisoning after Ingestion of Imported Jellyfish: Diagnostic Application of Serum Immunoassay." *Wilderness and Environmental Medicine* 6 (August): 288–294.

INDEX